TA
1675
.L363
1995

Laser experiments for
beginners.

$28.00

DATE			

BAKER & TAYLOR

LASER

EXPERIMENTS FOR BEGINNERS

LASER

EXPERIMENTS FOR BEGINNERS

Richard N. Zare*
Bertrand H. Spencer
Dwight S. Springer
Matthew P. Jacobson

*Department of Chemistry
Stanford University
Stanford, California

University Science Books
Sausalito, California

University Science Books
55 D Gate Five Road
Sausalito, CA 94965
Fax (415) 332-5393

Production Manager: *Stephen Peters*
Manuscript Editor: *Ann B. McGuire*
Designer: *Robert Ishi*
Illustrator: *John Choi*
Printer and Binder: *Maple-Vail Book Manufacturing Group*

This book is printed on acid-free paper.

Library of Congress Cataloging-in-Publication Data

Laser experiments for beginners / Richard N. Zare . . . [et al.].
    p.   cm.
   Includes bibliographical references and index.
   ISBN 0-935702-36-9
   1. Lasers —Experiments.   I. Zare, Richard N.
TA1675.L363   1995
621.36'6'078—dc20   94–43481
                                            CIP

Printed in the United States of America
10 9 8 7 6 5 4 3 2

Dedicated to our parents and our teachers.

CONTENTS

PREFACE

PURPOSE OF THIS BOOK

The laser is undoubtedly one of the most important inventions of the twentieth century. Although less than half a century has passed since their realization, lasers have been employed in a vast range of applications, including surgery, manufacturing, warfare, and telecommunications. Lasers have become particularly important in scientific research, especially in physics and chemistry, in which many discoveries would not have been possible without them. Although someday the laser may be considered one of the world's greatest inventions, along with the light bulb, the radio, and the computer, most people currently consider it esoteric and less practical for everyday life than these more familiar inventions. Many people know that lasers can be used to read compact disks and bar codes, but few know how a laser works or what scientific uses can be made of it.

This book introduces the usefulness of the laser as a tool in scientific research at a level of sophistication appropriate for advanced high school students and beginning college students. The aim of the majority of the experiments is to demonstrate how to use lasers to gain insight into chemical systems, primarily by studying how light interacts with matter. Although the subjects covered in the text are likely most appropriate in a chemistry curriculum, much of the material may also be of interest to physics and biology teachers and their students.

The two primary uses of the experiments are as classroom demonstrations or project-based laboratories. By project-based laboratories, we mean activities in which students have a role in decision-making rather than passively follow a list of instructions. These projects are designed not to illustrate textbook principles but to introduce the students to the techniques and challenges of scientific research. The material in this book is not intended to constitute a course in itself; instead, it is meant to provide a rich choice for supplemental lecture demonstrations and hands-on, in-depth special projects.

The equipment necessary to perform the experiments is relatively inexpensive and easy to acquire. All of the experiments are designed for low-power, low-cost lasers, such as helium–neon (He–Ne) or diode lasers. These lasers are very safe if used properly and may cost less than $100 (1995 price). Please refer to Chapter 1 for further discussion of the equipment involved in performing the experiments.

INTENDED AUDIENCE

The more challenging experiments in this book could make excellent additions to college laboratory programs, especially for freshmen and sophomores, because they would allow the students to investigate interesting chemical phenomena and simultaneously find out firsthand why the laser is such a useful tool of research. Most of the experiments could easily fit into the college format of one to two 3- to 4-hour laboratory sessions.

At the high school level, the experiments would be most appropriate as student projects that supplement the established curriculum. Although most of the experiments in this book require more time than is traditionally allotted to high school laboratories, we believe that the time spent performing them would be worthwhile because the students can learn about the scientific method and the challenges of research while they become familiar with an important piece of technology. High school students could work in groups to perform one of the experiments over two or more class periods, or individual students could use the experiments as starting points for independent experimentation (such as the year-long research projects typically assigned to Advanced Placement chemistry students). Lasers are already well-established tools in the curricula of many undergraduate institutions (1,2). We hope that in the near future, lasers will be equally at home in the science curricula of secondary schools.

ORGANIZATION OF THE TEXT AND SUBJECT MATTER

The Introduction provides practical information on each of the experiments to help teachers decide which experiments are the most appropriate for their laboratory programs. Chapter 1 provides information on the equipment needed to perform the experiments, including lasers and light-detection equipment. It also contains important statements about safety and chemical disposal.

The experiments are grouped into five chapters, each of which contains a short introduction to the scientific principles underlying the experiments in that chapter. The format of each experiment is influenced by Shakhashiri's excellent series *Chemical Demonstrations* (3) and is organized as follows:

- Degree of Difficulty
- Materials

- Procedure
- Hazards and Precautions
- Disposal
- Discussion
- References

Chapter 2 covers light scattering from disordered systems; most of the experiments use laser-light scattering to study colloidal suspensions. Chapter 3 centers on diffraction (light scattering from ordered systems) and gives examples of how to use lasers to demonstrate diffraction effects. Chapter 4 describes how a laser can be used to detect refractive index changes and provides two sample applications of this use of the laser. Chapter 5 provides insight into the electronic structure of matter with a series of experiments that use laser-induced fluorescence and an additional experiment on the magnetic properties of molecules. Chapter 5 also includes a brief discussion of the principles behind the operation of lasers. Chapter 6 presents an introduction to photochemistry and the use of lasers to induce and monitor photochemical reactions. The book concludes with a safety appendix that lists the known health hazards and recommended handling for each chemical, experiment by experiment.

For the purposes of this book, students need understand for the most part only that the laser is an intense light source with an output that is a beam of light, pure in color. For an overview of the operation of lasers and their applications, please refer to the books by Evans (4), Wilson and Hawkes (5), Andrews (6), Hecht (7), and O'Shea, Callen, and Rhodes (8), or the collection of articles presented as "Lasers from the Ground Up" in the June 1982 issue of the *Journal of Chemical Education,* especially the articles by Coleman (9) and Crosley (10). The books by Sirohi (11), Kallard (12), and McComb (13), the pamphlet by Metrologic Instruments, Inc. (14), and the report of a workshop on advanced laboratory experiments using lasers (15) also provide background information on lasers along with laser projects and experiments. Sirohi (11) and Kallard (12) focus on optical and physical phenomena such as holography, whereas the Metrologic publication (14) enumerates "More than 101 Ways to Use a Laser." McComb's book (13) is a particularly useful resource for technological details about lasers and "do-it-yourself" instructions for constructing electronic devices involving lasers.

ACKNOWLEDGMENTS

The initial enthusiasm for producing this book came from Richard N. Zare, Marguerite Blake Wilbur Professor of Chemistry, Stanford University, who directed the project and who wrote and revised this book. The choice and development of the experiments came primarily from Bertrand H. Spencer,[1] who teaches

---

1. Present address: Pinewood High School, Los Altos, CA 94022.

high school chemistry and worked in Professor Zare's laboratory from 1985 to 1990. Several students assisted in testing these experiments, including Daniel M. Albritton, Don H. Chin, Brian Mason, Colleen M. Sheridan, Julie Tinklenberg, George A. Willman, and Vicki Yamazaki.

Joining Zare and Spencer was Dwight S. Springer, Associate Professor of Chemistry, United States Military Academy, West Point, New York, who spent his sabbatical year (1990–1991) at Stanford University organizing experimental material and preparing write-ups. These tasks were completed by Matthew P. Jacobson, then a Stanford undergraduate chemistry major who had also done substitute teaching in high school chemistry and is currently a graduate student in the Department of Chemistry, Massachusetts Institute of Technology.

Numerous individuals contributed to this project with their advice, comments, and discussions. We are particularly indebted to Edouard R. Kouyoumdjian, Senior Staff Chemist, Environmental Health and Safety, Stanford University, who helped us prepare the safety appendix. All illustrations were ably prepared by John C. Choi, Graphics Technical Specialist, Stanford University. The authors assume responsibility for the accuracy of the material in this book, and we would be grateful to receive suggestions for corrections and improvements.

This work was supported in part by the National Science Foundation under grant numbers NSF MDR 84-70336 (1985–87), NSF MDR 87-51200 (1988–89), NSF MDR 89-54662 (1989–91). It was also made possible by the use of unrestricted funds from the Dean of Humanities and Sciences, Stanford University.

REFERENCES

1. Steehler, J.K. *J. Chem. Ed.* 67 (1990) A37.
2. Steehler, J.K. *J. Chem. Ed.* 67 (1990) A65.
3. Shakhashiri, B.Z. *Chemical Demonstrations,* vol. 1–4. University of Wisconsin Press; Madison, WI: vol. 1 (1983), vol. 2 (1985), vol. 3 (1989), vol. 4 (1992).
4. Evans, D.K. *Laser Applications in Physical Chemistry.* Marcel Dekker; New York: 1989.
5. Wilson, J.W., Hawkes, J.F.B. *Lasers, Principles and Applications.* Prentice-Hall; Englewood Cliffs, NJ: 1987.
6. Andrews, D.L. *Lasers in Chemistry,* 2nd ed. Springer; Berlin: 1990.
7. Hecht, J. *The Laser Guidebook.* McGraw-Hill; New York: 1986.
8. O'Shea, D.C., Callen, W.R., Rhodes, W.T. *Introduction to Lasers and Their Applications.* Addison-Wesley; Reading, MA: 1978.
9. Coleman, W.F. *J. Chem. Ed.* 59 (1982) 441.
10. Crosley, D.R. *J. Chem. Ed.* 59 (1982) 446.
11. Sirohi, R.S. *A Course of Experiments with He–Ne Laser.* John Wiley; New York: 1985.
12. T. Kallard. *Exploring Laser Light.* Optosonic Press. Reprinted by American Association of Physics Teachers; 5110 Roanoke Pl., Ste. 101, College Park, MD 20740: 1977.
13. McComb, G. *The Laser Cookbook.* TAB Books; Blue Ridge Summit, PA: 1988.
14. Metrologic Instruments, Inc. *Laser Teaching Supplement.* Metrologic Instruments; Blackwood, NJ: 1993.
15. Parks, J.E., Feigerle, C.S., Deserio, R., Wiest, J. *J. Laser Appl.* 6 (1994) 115.

INTRODUCTION: CHOOSING AN EXPERIMENT

This section provides information to help teachers decide which of the experiments in this book are the most appropriate for their laboratory programs. Table 1 provides an overview of each experiment's level of difficulty and most appropriate format (demonstration, in-class experiment, or project-based experiment). The text below explains how to interpret the table and addresses such practical concerns as

- Will my students be able to understand the experiment?
- Can the experiment be performed in one laboratory period?
- How much background do the experiments assume?
- Will the students need to prepare beforehand so they can understand the experiment?

Each experiment description includes a short synopsis at the beginning and a discussion section at the end. For specific information on the equipment necessary to perform the experiments (such as light-detection equipment), please refer to Chapter 1.

DEGREE OF DIFFICULTY

In Table 1, the classifications for degree of difficulty provide a guide to the appropriate level of students for each experiment. The level of experimental difficulty provides an indication of the sophistication of the experimental techniques used and, to some extent, the complexity (and cost) of the equipment needed. The categories are explained as follows:

Experimental Difficulty	Characteristics
Easy	Step-by-step, "cookbook" style instructions; assumes rudimentary laboratory skills
Moderate	Assumes proficiency in basic laboratory techniques, such as preparing solutions or using a separation funnel
Difficult	Assumes significant laboratory experience; requires patience and careful technique

TABLE 1. EXPERIMENTS AT A GLANCE

	Degree Of Difficulty		Principal Use		
	Experimental	Conceptual	Demo	In class	Project based
Chapter 2: Light Scattering from Disordered Systems					
2-1: The Tyndall Effect	Easy	Easy	X	X	
2-2: Dew Point Temperature and Cloud Formation	Easy	Easy	X	X	
2-3: Spatial Distribution and Polarization of Scattered Light	Easy	Moderate	X	X	X
2-4: Aggregation of Latex Microspheres	Moderate	Moderate		X	X
2-5: Emulsions and Microemulsions	Moderate	Moderate		X	X
2-6: Dynamics of the Gel-Formation Process	Moderate	Moderate			X
2-7: The Isoelectric Point of Gelatin	Moderate	Moderate			X
2-8: Scattering of Light by Optically Active Colloids	Moderate	Moderate	X		
2-9: Rate of Vesicle Aggregation	Difficult	Moderate			X
2-10: Light Scattering and Particle Size	Difficult	Difficult			X
Chapter 3: Diffraction: Light Scattering from Ordered Systems					
3-1: Diffraction from a Vernier Caliper	Easy	Moderate	X	X	
3-2: Optical Transforms	Easy	Moderate	X	X	X
3-3: Diffraction from a Two-Dimensional Crystal	Moderate	Moderate	X		X
3-4: Kossel Ring Diffraction Patterns from a Colloidal Crystal	Very difficult	Difficult	X		X
Chapter 4: Refraction of Light					
4-1: Column Chromatography with a Laser-Based Refractive Index Detection System	Moderate	Moderate			X
4-2: Extraction and Purification of Limonene	Moderate	Moderate			X
4-3: Synthesis and Purification of Oil of Wintergreen	Moderate	Moderate			X
4-4: Kinetics of the Acid-Catalyzed Conversion of Glycidol to Glycerol	Moderate	Moderate to difficult			X
Chapter 5: The Electronic Structure of Matter					
5-1: Extraction of Chlorophyll from Fresh Spinach	Moderate	Easy			X
5-2: Fluorescence vs. Concentration	Moderate	Moderate			X
5-3: Fluorescence Quenching	Moderate	Moderate			X
5-4: Fluorescence Depolarization	Moderate	Moderate			X
5-5: Magnetic Susceptibility Measurements	Difficult	Difficult	X		X

TABLE 1.   (*continued*)

	Degree Of Difficulty		Principal Use		
	Experimental	Conceptual	Demo	In class	Project based
Chapter 6: Photochemistry					
6-1: Photobleaching of Methylene Blue	Easy	Moderate	X	X	
6-2: Photoisomerization of Dimethylmaleate	Easy	Moderate	X	X	X
6-3: Photoinduced Polymerization of Acrylamide	Easy	Moderate	X	X	
6-4: Photooxidation of Diphenylisobenzofuran	Moderate	Moderate	X		X
6-5: Photochromism of Mercury Dithizonate	Difficult	Moderate	X		X
6-6: Ruthenium-Catalyzed Photoreduction of Paraquat	Difficult	Difficult			X

Experiments rated difficult may be inappropriate for the high school setting because of the complexity of the procedure.

The level of conceptual difficulty reflects the level of scientific background needed to understand the results of the experiment. Our meanings are as follows:

Conceptual Difficulty	Characteristics
Easy	Can be understood intuitively
Moderate	Requires the understanding of some background information, found in the discussion and introduction sections to each chapter
Difficult	Assumes a strong scientific background (i.e., a strong introductory course in chemistry, physics, or both); the information necessary to understand the experiment may be complicated or involve more sophisticated mathematical arguments

Experiments rated difficult conceptually are appropriate for beginning college students or the most advanced (and self-motivated) high school students.

PRINCIPAL USES OF THE EXPERIMENTS

One potential use of many of the experiments is as a **demonstration,** in which the teacher performs the experiment in front of a class to illustrate a scientific principle. An advantage of using the experiments this way is that the functions of sophisticated scientific instruments can be presented to students who might not yet be ready to perform the experiments themselves. A disadvantage is that the students do not get hands-on experience with the laser. Some of the demonstrations would likely require an entire class period to perform and explain.

Most of the experiments in this book are designed to be **project-based experiments.** Project-based experiments generally require more time than those designed to be performed within one or two high school class periods. These experiments typically require at least one hour and a half to complete.

Many of the project-based experiments would make excellent additions to college-level laboratory programs, especially for freshmen and sophomores. Most of these experiments could easily fit into the college format of one or two 3- to 4-hour laboratory sessions. The degree of difficulty ranges from moderate to difficult.

The project-based experiments can be integrated into the high school science curriculum in several ways. First, small groups of students could work on one project over two to three class periods. Although the time constraints involved in teaching high school chemistry are severe, we believe that the time spent performing project-based experiments is worthwhile because the students can learn about the scientific method and the challenges of research while they become familiar with the importance of the laser to scientific research. Second, many of these projects would be ideal for high school students required to perform an experiment individually for an end-of-the-year project (such as Advanced Placement chemistry students). Many projects give suggestions for further experimentation, and all could serve as starting points for independent student projects. We highly encourage students and teachers alike to use the experiments as starting points for further investigation of the subjects covered in this book.

Few of the experiments are appropriate for the traditional high school laboratory format in which students work in partners on an experiment designed to take one or possibly two class periods. The biggest constraint is that most high schools own only one laser, and all of the students could not have access to the laser simultaneously. The experiments that can be performed within two class periods are classified in Table 1 as **in-class experiments.**

Note that many of the experiments fit the criteria for more than one classification. Those classified as both in-class and project-based are best considered project-based experiments that could be performed in a relatively short time. Although these experiments could be done in some fashion in less than two hours, additional time would allow for more rigorous or further experimentation.

LASER

EXPERIMENTS FOR BEGINNERS

AN IMPORTANT NOTE TO THE READER

Laser and chemicals can be extremely dangerous if they are not handled properly and with care. Neither the publisher nor the authors assume any responsibility for injury or damages resulting from the unsafe application of the experiments discussed herein, or the unsafe operation of any lasers.

Safety and disposal information is provided with each experiment and compiled in an appendix at the end of this book. The reader must assume all responsibility to devise a safety plan for each experiment conducted.

CHAPTER 1

EQUIPMENT, SAFETY, AND DISPOSAL

This chapter provides practical information on lasers and other equipment necessary to perform the experiments in this book. It contains information relevant to all of the experiments and should be read before any experiments are performed. This information is presented only here and is not repeated in the individual experiments to which it might apply. Because the equipment in different science laboratories varies widely, the following statements provide broad guidelines instead of step-by-step instructions for how to conduct the experiments in this book. Particularly important are the sections on safety and disposal. Although safety and disposal information is provided with each experiment and collected in the Appendix, general comments are provided in this chapter. **The reader must assume the responsibility to devise a safety plan for each experiment conducted.**

EQUIPMENT AND SUPPLIES

Lasers

The experiments in this book are designed for low-power, low-cost lasers, such as He–Ne or diode lasers. We recommend the use of class 2 lasers, which are those that have an output power of less than 1 mW. When used properly, class 2 lasers are easy to operate safely in a laboratory setting. Please refer to the safety section later in this chapter for laser safety rules.

Most of the experiments were designed for a red He–Ne laser with an output wavelength of 632.8 nm or a red diode laser with an output wavelength of 670 nm. Several experiments require a laser with a wavelength of very nearly 633 nm. These are experiments 5-1 through 5-4, 6-1, 6-3, 6-4, and 6-6. For the remaining experiments, the output wavelength used is not crucial, although low-power red lasers are likely the least-expensive option.

For several experiments, having two lasers with different output wavelengths enhances the project, although performing these experiments with two different lasers is optional. These are experiments 2-3, 2-8, 2-10, 3-1 through 3-4, 6-2, 6-3, and 6-5. One relatively inexpensive option for a second laser is a He–Ne laser with an output wavelength of 543.5 nm, in the green region of the spectrum.

Lasers can be purchased from various suppliers (Table 1-1). The following is a list of some laser and laser-pointer models, their approximate 1995 cost, and a supplier (in parentheses):

1. Minilaser SE-9367 $270.00 (Pasco)

2. Laser Pointer SE-9716 $97.00 (Pasco)

3. Laser Pointer 3 mW $49.95 (Metrologic)

4. VDL Solid State Laser 1 mW MC-268 $250 (Metrologic)

5. He–Ne Laser 0.5 mW $215.00 (Edmund Scientific)

6. Laser Pointer 3 mW $49.95 (Edmund Scientific)

7. He–Ne Laser 0.5 mW 05LLR811 $205.00 (Melles Griot)

8. He–Ne Laser 0.5 mW U-1508 $315.00 (Newport)

9. Laser Pointer 2 mW 1 GN3 $68.00 (MWK Industries)

10. Laser Pointer 5mW P650 $120.00 (Beta Electronics)

TABLE 1-1.   EQUIPMENT AND CHEMICAL SUPPLIERS

Aldrich Chemical Co., P.O. Box 355, Milwaukee, WI 53201, 800/558-9160

Beta Electronics, Inc. (laser products), 2209 Cloverdale Ct., Columbus, OH 43235-2723, 800/546-2382

Cross Educational Software, 1802 N. Trenton St., Ruston, LA 71270, 318/255-8921

Edmund Scientific, 101 Gloucester Pike, Barrington, NJ 08007, 609/573-6295

Meridith Instruments, P.O. Box 1724, Glendale, AZ 85311, 602/934-9387

Metrologic Instruments, Inc., Coles Road at Rte. 42, Blackwood, NJ 08012, 800/IDMETRO

Melles Griot (high-quality laser products), 1770 Kettering St., Irvine, CA 92714, 800/835-2626

Millipore, 80 Ashby Rd., Bedford, MA 01730, 800/221-1975

MWK Industries, 1269 W. Pomona, Corona, CA 91720, 909/278-0563

Newport Corp., 18235 Mt. Baldy Circle, Fountain Valley, CA 92728, 714/963-9811

Oriel Corporation (optical components), 250 Long Beach Blvd., Stratford, CT 06497, 203/377-2550

Pasco Scientific, P.O. Box 619011, Roseville, CA 95661, 800/772-8700

Sargent-Welch, 1617 East Ball Road, Anaheim, CA 92803, 714/772-3550

Sigma Chemical Co., P.O. Box 14508, St. Louis, MO 63178, 800/325-3010

Thorlabs, P.O. Box 366, 75 Mill St., Newton, NJ 07860-0366, 201/579-7227

Vernier Software, 2920 Southwest 89th St., Portland, OR 97225, 503/297-5317

Surplus equipment vendors such as Meridith Instruments and MWK Industries often have lasers and parts at even lower cost. For example, MWK Industries offers a 0.5 mW He–Ne laser kit (which includes laser tube, power supply, and ballast resistor) at $48.00. Interested parties should check with the vendor for price, specifications, and availability. For the construction-minded reader, a solid state laser/photodetector system can be built from components that cost approximately $125.00 (see reference 1 for details).

The design of the He–Ne laser is discussed in the introduction to Chapter 5 and in most basic textbooks on lasers (2–5).

Light Detection

Several experiments in this book use lasers in ways that require no light-detection system; these are experiments 2-1, 2-3, 2-5, 2-8, 3-1 through 3-4, 4-4, 5-1, 6-1 through 6-3, and 6-5. For the remaining experiments, making quantitative measurements of light intensity is helpful or necessary.

Most sophisticated light-detection systems are costly, but affordable alternatives are available. These systems are either computer-based or non–computer-based. All the methods of detection used in this text, whether computerized or not, use the silicon phototransistor, which is very light sensitive. If the device becomes overloaded, positioning crossed optical polarizers or other light-blocking components (neutral density filters) in front of the detector may be necessary to reduce the amount of light entering the detector.

Computer-based light detection. Depending on the type(s) of computer present in the laboratory, the following equipment packages, including hardware and software, are recommended. Prices are approximate for 1995. All of the vendors (listed in parentheses) furnish ample documentation with easy-to-follow installation and operation instructions.

1. IBM (Vernier)
   Sensor LFDIN $39.00
   Voltage Input VIUIBM $45.00
   Game Port Card IGP $25.00
   Voltage Plotter System VPIBM $39.95

2. Macintosh (Vernier)
   Sensor LFDIN $39.00
   Interface Serial Box SBI $99.00
   Data Logger DL-SBI $30.00

3. Apple IIe (Cross Educational Software)
   Game Port Attachment and Software CR-130 $30.00

In addition, the Montana State University Chemistry Department (Bozeman, MT) has developed interfaces for both Apple and IBM computer systems that include light and temperature sensors as well as a laboratory manual. Schematic diagrams are also available for making A/D converters for the Apple (6) and IBM personal computers (7).

Light measurements taken on a computer-interfaced light-detection system can be stored on the computer hard disk or a floppy disk. For example, game port applications on the Apple IIe can be programmed to store data as a disk file and then retrieve the data by printing out the file. Software developed by Cross Educational Software (Ruston, LA) may be useful for storing data. Similarly, the IBM and Macintosh applications allow up to 24 hours of data collection at an accumulation rate of 50 sec^{-1}. Alternate equipment packages can extend this data storage if needed. For those knowledgeable in computer programming, an additional benefit of using a computer-interfaced system is that a few lines of code can be written to average light-intensity measurements over periods of time, thereby reducing the "noise" in the data.

Other light-detection equipment. If a computer is not available for laboratory use, a photometer or photodiode may be an affordable option for light detection. Several instruments and suppliers (in parentheses) are listed below for ready-to-use equipment:

1. High Sensitivity Photometer OS-8020 $540.00 (Pasco)

2. Student Photometer OS-9152B $275.00 (Pasco)

3. Digital Laser Power Meter 45-540 $439.00 (Metrologic)

4. Photometer 45-230 $179.00 (Metrologic)

A photometer can be used in experiments 2-2, 2-3, 4-1 through 4-3, 5-5, 6-5, and 6-6.

Several experiments require more sensitive measurements of light intensity. These are experiments 2-4, 2-6, 2-7, 2-9, 2-10, and 5-2 through 5-4. A computer-interfaced silicon photodiode is best for these experiments, but a sensitive photometer with fiber-optic capability (such as PASCO OS8020, $540) is also sufficient. The data from photometers can be recorded by hand or on a chart recorder if a continuous trace of light intensity is desired.

For well under $100, a silicon diode photodetector that operates off an internal 22.5-volt battery can be purchased from Thorlabs. To operate this instrument, attach a BNC TEE at the BNC connector end of the detector. Across one of the terminals of the TEE, attach a 1,000-ohm resistor and plug the other end into a multimeter via a BNC cable. Using the 1,000-ohm resistor, a voltage range of 0–1.000 VDC can be obtained for 0–3 mW of laser power. A larger resistance will increase the spread somewhat. This level of sensitivity is sufficient for most of the applications described in the text. The addition of an optical fiber attachment to the assembly would make this system the most attractive of the non–computer-based solutions for light detection. Mounting the detector requires an 8-32 threaded rod that can be clamped down to the optical table. To ease assembly of the working light detector, we include this brief list of parts and a schematic guide for use in assembly (Figure 1-1):

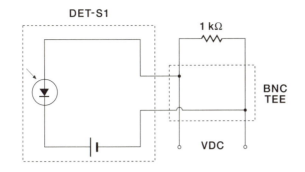

FIGURE 1-1. Schematic diagram of a Thorlabs light detector.

1. Detector DETI-S1 $78.00 (Thorlabs)

2. Battery replacement BAT22.5 $5.00 (Thorlabs)

3. Resistor 1,000-ohm (Radio Shack, others)

4. BNC connectors and cables (Radio Shack, others)

For even less money (~$35.00), a photodetector can be assembled from parts. Instructions for assembly can be found in reference 1 and page 49 of reference 5. The parts for a reverse-biased photodiode are available from Thorlabs, and the schematic is shown in Figure 1-2. The response of this device should not be that different from the Thorlabs DETI-S1, although mounting and housing the photodiode will not be as advantageous. The parts are as follows:

1. Photodiode Si PIN photodiode (for 670 nm) FDS100 $6.00 (Thorlabs)

2. Socket S8058 $2.80 (Thorlabs)

3. Riser Mount RB3-LH1 $8.00 (Thorlabs)

Batteries, resistor, connectors, and cables are as listed above for the DETI-S1.

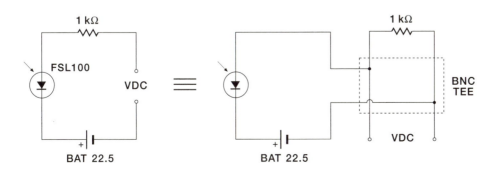

FIGURE 1-2. Schematic diagram of a component-based light detector.

Radio Shack and Tandy also sell silicon phototransistors that can be reversed biased with a 9-volt alkaline battery in series with a 1-megaohm resistor and read with a digital voltmeter. The performance of such components has not been verified by these authors.

Sample Cell Holders

In experiments that involve quantitative measurements of light intensity, slight movements of the sample cell can misalign the light-detection system and lead to large errors. A sample cell holder similar to the one shown in Figure 1-3 can be used to immobilize the sample cell. Note that the holder shown is designed for a standard 1-cm-square cuvette, which is used as a sample cell in

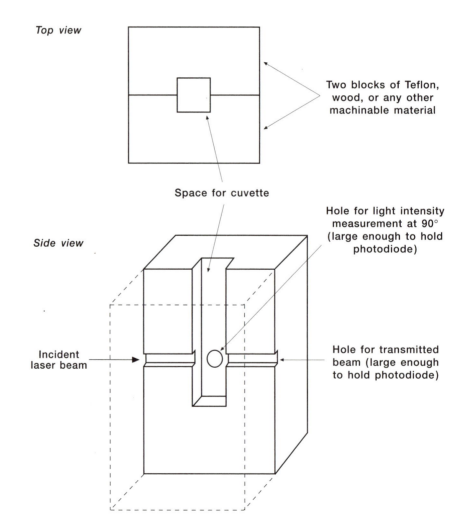

FIGURE 1-3. Schematic diagram of sample cell (cuvette) holder.

most of the experiments in this book. This basic design may need to be modified for experiments that employ polarizers or filters in the detection system. A sample cell holder is recommended in experiments 2-4, 2-6, 2-7, 2-9, 2-10, 5-2 through 5-4, and 6-6. In addition to minimizing movement of the sample cell, the sample cell holder also isolates the sample from room light. For all applications that involve the study of 90° scattered or fluorescent light using the sample holders, we recommend drilling a hole slightly larger than the diameter of the optical fiber of the light detector.

Beam Splitters

An important consideration in many of the quantitative experiments is that the power output of the lasers fluctuates over time. Diode lasers usually have smaller fluctuations than comparable He–Ne lasers. For both types of lasers, these fluctuations are approximately sinusoidal. For the gas lasers, the periods of fluctuation are short (a few seconds) when the laser is just turned on and lengthen once the laser has "stabilized."

One way of correcting for these fluctuations is to split the laser beam into two components: a probe beam and a reference beam. The probe beam enters the sample and is used to make the desired measurements on the system, while the reference beam is directed at a light meter to measure the laser's output power. All measurements can then be taken when the laser power is at some fixed value, so that the measurements do not need to be corrected for the oscillations. The key element in this procedure is a beam splitter, which divides the incoming ray into a refracted beam and a transmitted beam at right angles to each other. Beam splitters are available commercially, but a plate of glass can serve as a simple beam splitter because it transmits most of the light but reflects a small portion of it (Figure 1-4).

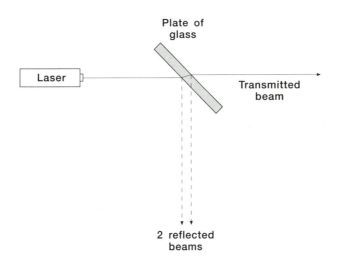

FIGURE 1-4. Splitting of laser beam into reflected and transmitted beams by a plate of glass.

Beam splitters are recommended in experiments 2-6, 2-7, 2-9, 2-10, 5-2 through 5-4, and 6-6. Note that the use of a beam splitter requires the use of an extra light detector to monitor the power output from the laser. This light detector does not need to be particularly sensitive, however.

A complete line of moderately priced optical equipment such as gratings, lenses, mirrors, splitters, and mounts is available from Edmund Scientific (Table 1-1).

Water Filters

Several experiments in this book involve directing the laser beam through an aqueous solution and monitoring the intensity of light at right angles to the laser beam. Aqueous solutions often contain a large amount of dust, which can scatter the light from the laser at right angles to the laser beam and thus interfere with the measurements being taken. A simple solution to this problem is to pass the water used to make the solutions (or the final solutions themselves) through filters with very small pore spacings, which are available from suppliers such as Millipore or Sargent-Welch at low cost (Table 1-1). A 0.22-μm filter removes virtually all dust particles, but a 0.45-μm filter suffices for many of the experiments. Please refer to the procedures of the individual experiments for which type of filter to use. The filters are available in various designs; those designed to be attached to a syringe are versatile and economical. The experiments that recommend use of a water filter are experiments 2-4, 2-6, 2-7, 2-10, 5-2, and 5-3.

Purchasing Chemicals

Most of the chemicals used in the experiments are relatively inexpensive, but many are not commonly found in educational laboratories. For this reason, we have included at least one source for each of the more obscure chemicals, usually either Sigma or Aldrich, because these companies carry a particularly wide variety of compounds (Table 1-1). We do not, however, endorse any particular company over another and recommend readers consult more than one catalog to find the lowest price and most appropriate quantity of the desired compound.

SAFETY

The American National Standards Institute has established safety rules and ratings for lasers (8–10). Lasers with a power of less than 1 mW are classified as class 2 lasers and are the most appropriate for use in the teaching laboratory. **The basic safety rule is to avoid looking directly into the laser beam.** Although 1 mW seems small compared with a 100-W lightbulb, all of the energy is concentrated to a roughly 1-mm^2 spot, making the energy per area very large.

Because the eye can focus the already intense laser beam onto a small area of the retina, permanent damage can result from extended viewing of the direct beam. An accidental glance, however, is no more likely to cause instantaneous blindness than is an accidental glance at the sun (3). In addition, the eye becomes sore with prolonged viewing of diffuse or reflected light. Set up all experiments to minimize the chances of any such exposure!

Lasers are classified according to their capability to produce injury to personnel, and this classification depends both on the wavelength and the intensity of the emitted radiation. Class 1 lasers denote lasers that cannot produce a hazard under normal operating conditions. Class 2a lasers are low-power visible lasers that are not intended for prolonged viewing; under normal operating conditions, they will not produce a hazard if viewed directly for periods not exceeding 1,000 seconds. Class 2 lasers denote low-power visible lasers that, because of the normal human aversion response, do not normally present a hazard but may if viewed directly for extended periods of time. Class 3a lasers are lasers that would not produce a hazard if viewed only briefly with the unprotected eye. They may present a hazard if viewed using light-collecting optics. Class 3b lasers denote lasers that can produce a hazard if viewed directly. Class 4 lasers can produce a hazard not only from direct viewing or a specular reflection but also from diffuse reflection. Class 2 lasers present no danger to the skin, and the beam does not even feel warm on the skin. Use of class 3 and class 4 lasers requires protective eye goggles and other safety precautions. These lasers are generally too high powered and expensive for use by beginning students.

In addition to the precautions necessary for operating lasers, **all of the safety rules and procedures applicable to chemistry laboratories must be also applied, including the use of safety goggles and the proper storage, handling, and disposal of chemicals.** To provide guidance, we have collected in the Appendix the National Fire Protection Association's hazard classification, health hazard description, and recommended handling of each compound used, listed experiment by experiment.

All work should be performed under the supervision of a teacher qualified to supervise a chemistry laboratory. As a special note to high school teachers, the experiments in this book are designed with simplicity in mind, but some of the chemicals used, such as perchloric acid or heptane, are not typically used in high school chemistry laboratories. Special attention should be given to the use of such chemicals. In each experimental description, we have attempted to identify relevant hazard information for any potentially problematic substance. Materials Safety Data Sheets (available from chemical manufacturers and reference 11) are reliable sources of safety information on specific chemicals. Other good sources of safety information are *The Merck Index* (12) and the American Chemical Society publication *Safety in Academic Laboratories* (13). Still other sources that may be valuable in designing a safe laboratory are included as references 14 through 21.

DISPOSAL OF CHEMICALS

As recently as the 1970s, the most common method of disposing of chemicals was to dump them down the drain. Today this practice is no longer acceptable with any but the most benign chemical compounds. Chemical disposal regulations vary widely from region to region. Federal, state, local, city, and water district regulations all govern the disposal of chemicals; thus, no generally applicable guidelines can be set for how to dispose of most of the chemicals used in this book. We do, however, indicate which chemicals can be disposed of down a laboratory drain and which chemicals need to be treated in accordance with disposal regulations. We also point out compounds known to be particularly hazardous waste and that require special treatment procedures. A resource that some teachers may find valuable is the book on chemical disposal published by the National Research Council (22), although some of the information is now out of date. In all cases, the experimenter holds the responsibility to dispose of chemicals properly, and the authors and publishers of this book assume no responsibility or liability for the use of any chemical or procedure specified in this book.

REFERENCES

1. Sieflin, H.E., Tomaschke, H.E. *Phys. Teach.* 32 (1994) 315–317.
2. Andrews, D.L. *Lasers in Chemistry,* 2nd ed. Springer; Berlin: 1990.
3. Hecht, J. *The Laser Guidebook,* McGraw-Hill; New York: 1986.
4. Sirohi, R.S. *A Course of Experiments with He–Ne Laser.* John Wiley; New York: 1985.
5. McComb, G. *The Laser Cookbook.* TAB Books; Blue Ridge Summit, PA: 1988.
6. Wisman, R., Forinash, K. *Am. J. Phys.* 59 (1991) 187.
7. Leiker, P.S. *Phys. Teach.* 28 (1990) 332.
8. *Laser Safety Reference Book.* Laser Institute of America; Toledo: 1985.
9. Winburn, D.C. *Practical Laser Safety.* Marcel Dekker; New York: 1985.
10. Wilson, J.W., Hawkes, J.F.B. *Lasers, Principles and Applications.* Prentice-Hall; Englewood Cliffs, NJ: 1987; 298–301.
11. *Materials Safety Data Sheets.* CHEMTREC (a division of the Chemical Manufacturers Association, 2501 M Street, NW, Washington, DC 20037).
12. Budavari, S. (ed). *The Merck Index: An Encyclopedia of Chemicals, Drugs, and Biologicals,* 10th ed. Merck and Co.; Rahway, NJ: 1989.
13. The American Chemical Society Committee on Chemical Safety. *Safety in Academic Chemistry Laboratories,* 4th ed. American Chemical Society; Washington, DC: 1985.
14. Furr, A.K. *CRC Handbook of Laboratory Safety,* 3rd ed. CRC Press; Boca Raton, FL: 1990.
15. Dux, J.P., Stalzer, R.F. *Managing Safety in the Chemical Laboratory.* Van Nostrand Reinhold; New York: 1988.
16. Ashbrook, P.C., Renfrew, M.M. *Safe Laboratories, Principles and Practices for Design and Remodeling.* Lewis; Chelsea, MI: 1991.
17. The Royal Society of Chemistry. *Safe Practices in Chemical Laboratories.* The Royal Society of Chemistry; London: 1989.
18. The Royal Society of Chemistry. *Health and Safety in the Chemical Laboratory. Where Do We Go from Here?* The Royal Society of Chemistry; London: 1984.

19. Bretherick, L., (ed). *Hazards in the Chemical Laboratory,* 4th ed. The Royal Society of Chemistry; London: 1986.

20. Lewis, R.J. Sr. *Hazardous Chemicals Desk Reference,* 2nd ed. Van Nostrand Reinhold; New York: 1991.

21. Lenga, R.E. *The Sigma-Aldrich Library of Chemical Safety Data,* 2nd ed., vols. 1 & 2. Sigma-Aldrich; Milwaukee: 1988.

22. Committee on Hazardous Substances in the Laboratory. *Prudent Practices for Disposal of Chemicals from Laboratories.* National Research Council; Washington, DC: 1983.

CHAPTER 2

LIGHT SCATTERING FROM DISORDERED SYSTEMS

One way light can interact with matter is through the process of light scattering, in which particles in the path of a beam of light disperse the light in many directions. This chapter presents 10 experiments that use a laser to study light scattering. Two of the experiments (2-3 and 2-10) examine the theory of light scattering, and the rest use light scattering as a tool to investigate the properties of dispersions known as colloids, which contain particles much too small to be seen.

In the introduction below, we present a simple model of light scattering, which is useful for interpreting the results of some of the chapter's experiments and for explaining everyday phenomena such as why the sky is blue. The information in the subsequent section, "Volume and Wavelength Dependence of the Intensity of Scattered Light," is necessary to understand experiment 2-10 and, to a lesser extent, experiment 2-7. In addition, Experiments 2-3, 2-8, and 2-10 rely on an understanding of the topics presented under "Spatial Distribution and Polarization of Scattered Light." The remainder of the experiments do not require an understanding of the theory of light scattering, although an understanding of the physical basis of light scattering is helpful. The information in the final section, "Light-Scattering Experiments," is important for all of the experiments.

INTRODUCTION TO LIGHT SCATTERING

The study of light scattering dates to the early eleventh century, when the Arabic scientist Alhazen of Basra tried to explain why the sky is blue. Alhazen did not use the term *light scattering* but rather accounted for blue skies by a preferential "reflection" of sunlight by particles contained in air. Although this explanation is not entirely correct, Alhazen was the first to explain the blueness of the sky in terms of an interaction between light and matter.

Centuries later, European scientists, such as Leonardo da Vinci, investigated the same question of why the sky is blue by performing experiments with particle-containing gases that scattered light in much the same way as the Earth's atmosphere. Despite these pioneering studies, da Vinci wrote: "Experience shows us that air must have darkness beyond it and hence it appears blue." By the nineteenth century, however, light scattering experiments had revealed the following observations:

1. Larger particles scatter light more effectively than smaller ones.

2. Small particles scatter higher frequency light more than they scatter lower frequency light (e.g., small particles scatter blue light more effectively than red light).

3. Light scattered at 90° to an incident, unpolarized beam is often highly polarized.

In the late nineteenth century, Lord Rayleigh provided a model of light scattering that accounted for these observations. In accordance with classical electromagnetic theory, Rayleigh treated a beam of light as an oscillating electromagnetic wave that can interact with particles by causing the negatively charged electrons in the particle to oscillate at the same frequency as the incident light. The particle, which can be as small as a molecule or atom, then behaves like an oscillating electric dipole, which reradiates electromagnetic energy at the same frequency as the incident light.

VOLUME AND WAVELENGTH DEPENDENCE OF THE INTENSITY OF SCATTERED LIGHT

Lord Rayleigh provided a qualitative argument based on dimensional analysis to model the total intensity of scattered light for the hypothetical case of a very small sphere. This argument is somewhat involved and may not be appropriate for beginning science students. The end result of the argument is equation 2-3, which is the basis for understanding the results of experiments 2-1, 2-2, 2-7, and 2-10.

The starting point of Rayleigh's argument is the assumption that the intensity of scattered light from a particle should depend on the size of the particle (corresponding to a volume V), the distance r from the particle to the observer, the wavelength λ of the scattered light, and the refractive indices n_1 of the particle and n_2 of the environment. Mathematically, this condition can be expressed as

$$I = f(V, r, \lambda, n_1, n_2) I_o \qquad (2\text{-}1)$$

where I_o is the intensity of the incident light, I is the total intensity of scattered light, and the f is a dimensionless function of five variables (1).

Rayleigh reasoned that because n_1 and n_2 are dimensionless, the dimensions of V, r, and λ must cancel each other to make f dimensionless. To specify the dependence of I on the three variables V, r, and λ, Rayleigh made the important assumption that the particle is so small that it can be approximated as a point source. Physically, this approximation implies that the particle has dimensions much smaller than the wavelength of incident light, and Rayleigh theory accurately models light scattering only when this size relationship is the case.

For small particles that behave as point sources, classical electromagnetic theory predicts the dependence of I on the volume V and the distance r from the particle to the observer. For a point source, the intensity of radiation falls off as $1/r^2$, corresponding to units of $(\text{length})^{-2}$. Further, for a very small particle, the intensity of scattered light should be proportional to the square of the volume of the particle, because the electric field (dipole moment) induced in the particle should be proportional to the volume of the particle. The intensity of the reradiated light should be proportional to the square of the electric field strength of the dipole, so I is proportional to V^2, which has the units of $(\text{length})^6$. Thus, for the function f to be dimensionless, Lord Rayleigh concluded that the dependence of I on λ must have the units of $(\text{length})^{-4}$, which implies that intensity of scattered light varies as $1/\lambda^4$ (2,3). These arguments lead to the following formula for the intensity of light scattered from a small sphere:

$$I = \left(\frac{V^2}{r^2 \lambda^4}\right) I_o \, f(n_1, n_2) \qquad (2\text{-}2)$$

where f is some dimensionless function of the refractive indices of the particle and the environment.

When more than one scattering center are present, constructive and destructive interference can occur between the light scattered from the different particles (see Chapter 3 for a discussion of interference). If the scattering centers have a regular spatial order, as in a crystal, then the scattered light tends to cancel itself in almost all directions except in the direction of the incident light beam. This phenomenon is known as *diffraction* and is the subject of the next chapter. By contrast, the molecules in a gas are disordered, and microscopic density fluctuations in the gas ensure that the constructive and destructive interference of the light scattered from all of the different particles tends to cancel itself. In fact, for most disordered systems, the total intensity of scattered light can be shown to be equal to the sum of the intensities of light scattered from each particle. If so, then in Rayleigh's model, the intensity of scattered light from N small spheres with volume V is

$$I = \left(\frac{N \, V^2}{r^2 \, \lambda^4}\right) I_o \, f(n_1, n_2) \qquad (2\text{-}3)$$

Note that the intensity of scattered light is proportional to the number of scattering particles, which should be proportional to the concentration of scattering particles in any given sample.

Rayleigh's model of scattering helps explain why the sky is blue. When we look at the sky, we observe light that has been scattered off particles in the atmosphere, such as the gas molecules of which the atmosphere is composed. The major components of the atmosphere, nitrogen (N_2), oxygen (O_2), and water (H_2O) molecules, have dimensions much smaller than the wavelengths of visible light, and thus we expect that Rayleigh theory should accurately model atmospheric scattering. Since the intensity of scattered light (I) is inversely related to the fourth power of the wavelength of incident light,

$$I \propto (1/\lambda)^4 \qquad (2\text{-}4)$$

blue light should scatter more than red light. We can compute the ratio of the
intensity of scattered blue light (450 nm) to the intensity of scattered red light
(650 nm) to be roughly

$$\frac{I_b}{I_r} = \frac{(1/\lambda_b)^4}{(1/\lambda_r)^4} = \left(\frac{\lambda_r}{\lambda_b}\right)^4 = \left(\frac{650 \text{ nm}}{450 \text{ nm}}\right)^4 = 4.4 \tag{2-5}$$

Note, however, that purple (violet) light has a shorter wavelength than blue light
and should scatter even more effectively. This prediction is confirmed experimentally, as shown in Figure 2-1. Why not a purple sky rather than a blue one?
A complete explanation for the perceived color involves two additional factors:
the nature of the solar spectrum and the spectral response of the human eye (4).
The solar spectrum is depicted in Figure 2-2. Note the modest peak in the intensity of the visible solar spectrum above the atmosphere in the range of 460 to
480 nm, which corresponds to blue light. In addition, Figure 2-3 shows that the
eye's sensitivity to visible light peaks toward wavelengths longer than blue. This
combination of factors results in the appearance of the sky as blue to the human
eye.[1]

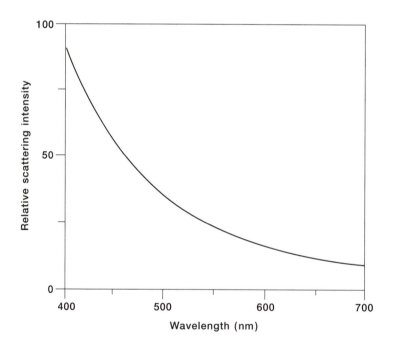

FIGURE 2-1. Wavelength dependence of scattering by atmospheric molecules.

1. The blue color of water also has an interesting explanation that is related to overtone vibrational absorptions in the red region of the spectrum. A complete discussion is found in reference
5.

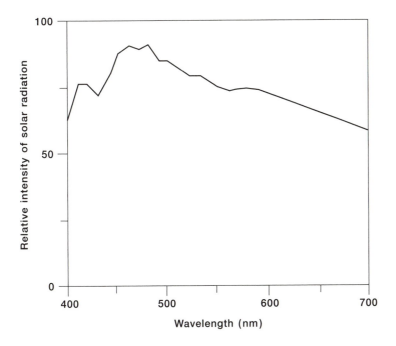

FIGURE 2-2. Intensity of visible solar radiation outside the Earth's atmosphere.

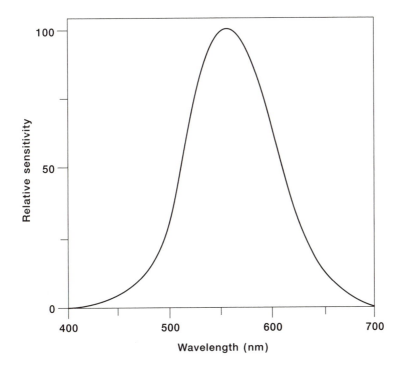

FIGURE 2-3. Sensitivity of the human eye to visible light.

We can also explain sunsets in terms of scattering. Because blue light is scattered more than red light, when we look toward the sun at sunset we see more longer wavelengths than shorter wavelengths, that is, a preference for yellows, oranges, and reds. The beautiful red sunsets seen after volcanic eruptions or in many large cities are a result of atmospheric particulates. The particles in the atmosphere scatter not only the blue and green colors but also yellows and oranges, thus shifting the transmitted solar spectrum into the red. The beautiful pink colors seen in clouds after sunset stem from scattered light that originates beneath the horizon and reflects off clouds. Additional discussions of these concepts can be found in references 4 and 6.

SPATIAL DISTRIBUTION AND POLARIZATION OF SCATTERED LIGHT

Thus far, we have considered only the total intensity of light scattered from very small particles. We turn to the expected spatial distribution and polarization of the scattered light. Consider the electric field of a beam of polarized light incident on a scattering sample (the magnetic field will behave in the same manner). The dipoles induced in the scattering particles oscillate with the same relative polarization as the incident light beam; that is, the oscillations induced are parallel to the electric field of the incoming light and perpendicular to the direction of propagation (Figure 2-4). The oscillating dipoles then reradiate the light with the same polarization.

Electromagnetic theory tells us that the intensity of a beam of light is proportional to the square of the electric field amplitude. This theory allows us to predict the intensities of scattered light at any angle with respect to the dipole. Specifically, we can calculate the intensity of scattered light by considering the amplitude of the scattered radiation. Imagine placing a light detector at various positions around the dipole. The amplitude of the reradiated light observed by the light detector at any position should be proportional to the projection of the dipole moment on a plane perpendicular to the viewing axis, as shown in Figure 2-5. Thus, the amplitude of the reradiated light is at a maximum in any direction

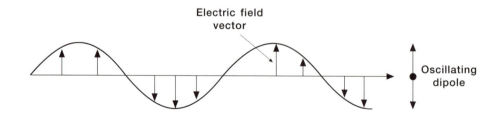

FIGURE 2-4. Polarized light beam incident on a particle.

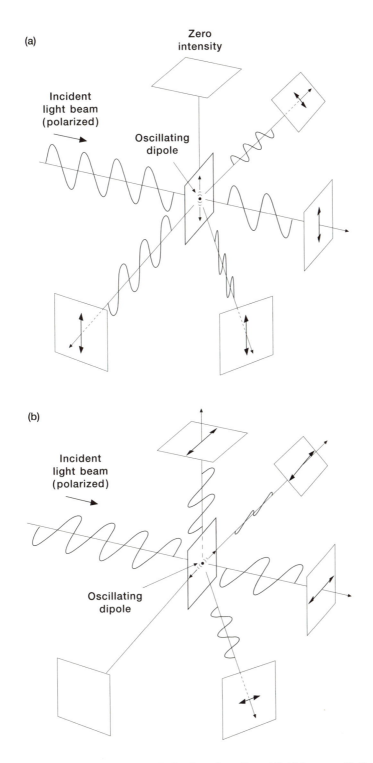

FIGURE 2-5. Spatial distribution and polarization of scattered light from oscillating dipoles induced by plane polarized light with its electric field pointing **(a)** in the vertical direction and **(b)** in the horizontal direction.

perpendicular to the dipole axis. Conversely, the amplitude of the reradiated light is zero in the direction of the dipole axis, and virtually no scattered light intensity should be observed in this direction. At any intermediate position, the amplitude varies as sin Φ, where Φ is defined as the angle between the line of sight and the dipole axis (Figure 2-6), and the intensity of light varies as sin^2 Φ.

Consider what happens when the incident wave consists of unpolarized light. The dipoles induced in the scattering particles oscillate in any direction perpendicular to the beam of incident light. In this case, the intensity of scattered light can be considered a superposition of many different dipoles such as the ones in Figure 2-5, oriented at random directions perpendicular to the incident beam. With this model in mind, the light scattered in the forward direction (the same direction as the light beam) should be completely unpolarized (Figure 2-7), whereas the light should be completely linearly polarized in all directions perpendicular to the incident beam. At any positions off axis, the light is partially

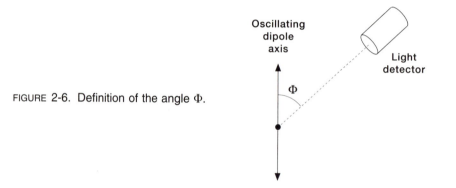

FIGURE 2-6. Definition of the angle Φ.

polarized. This symmetric variation in the degree of polarization with scattering angle θ (which is defined as the angle between the line of sight and the incident beam) is shown in Figure 2-8. The intensity of the scattered light can be modeled as

$$I \propto I_o \left(1 + \cos^2 \theta\right) \qquad (2-6)$$

Note that the intensity peaks at 0° and 180° but that there are no regions of zero intensity. The light scattered at 90° should be one-half as intense as that at 180° and should be completely plane polarized.

An experimental confirmation of this last prediction is that the blue light scattered at right angles to the sun is significantly polarized. You can confirm this by looking at the sky through a sheet of Polaroid at 90° to the sun. The light intensity you observe will vary considerably as the polarizer is rotated. The degree of plane polarization is incomplete because Rayleigh theory only approximates the scattering conditions in the atmosphere. Rayleigh theory does not account for multiple scattering, in which more than one scattering event occurs before observation. Further, some particles in the atmosphere are too

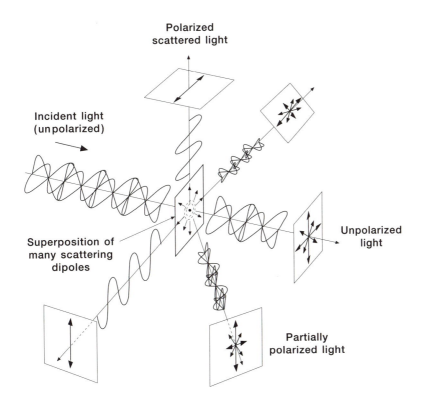

FIGURE 2-7. Spatial distribution and polarization of scattered light from oscillating dipoles induced by unpolarized light.

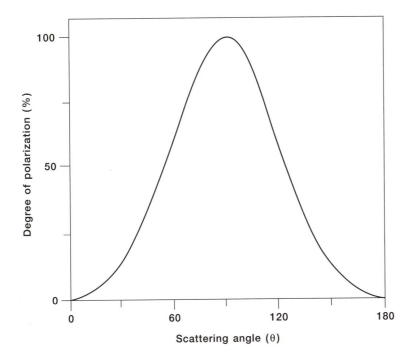

FIGURE 2-8. Spatial dependence of the polarization of scattered light when the incident beam is unpolarized.

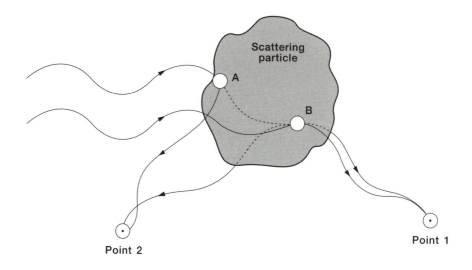

FIGURE 2-9. Interference of light scattered from two different points, A and B, on a scattering particle. At point 1, the waves interfere constructively; at point 2, they interfere destructively.

large to be treated by Rayleigh theory or cannot be accurately approximated as uniform polarizable spheres (molecular anisotropy). Molecular anisotropy is believed to be the principal cause of the incomplete polarization of scattered skylight (7).

Several more complex theories of light scattering can be used to model light scattering from particles too large to be accurately described by Rayleigh theory. Among other factors, these theories take into account that the light scattered from one section of a particle may become out of phase with that scattered from another section, resulting in interference (Figure 2-9). In this case, the scattered intensity is not symmetric about the particle, as predicted by Rayleigh theory (Figure 2-10). One theory that accounts for particle size, shape, dielectric constant, and absorbance is Mie theory, named after Gustav Mie (1868–1957) (8,9). An important prediction of Mie theory is that as the particle size increases, the wavelength dependence of the intensity of the scattered light decreases, so that the Rayleigh $1/\lambda^4$ dependence no longer holds. This prediction helps explain why clouds are white. The water droplets in clouds are generally much larger than the wavelengths of visible light, and thus we would not expect Rayleigh theory to accurately model the scattering from clouds. In particular, the wavelength dependence of the scattered light intensity is sharply decreased, and a broad spectrum of light is scattered off clouds, making them appear white to us.

LIGHT-SCATTERING EXPERIMENTS

The experiments in this chapter use laser light scattering as a tool to probe various properties of chemical systems. Lasers are particularly useful for light-scattering experiments because scattered light is generally quite weak, and an

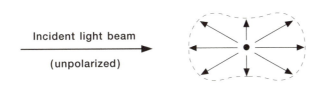

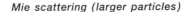

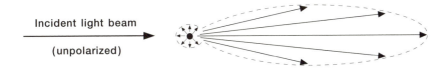

FIGURE 2-10. Comparison of Rayleigh and Mie scattering. Rayleigh theory is accurate in the limit of small particles and predicts that the intensity of scattered light is symmetric about the particle. Mie theory applies to larger particles and predicts that the intensity of scattered light is asymmetric about the particle.

intense light source such as a laser allows for more sensitive measurements of scattering intensity.

Desirable samples for scattering experiments are ones in which the scattering particles are large enough to produce significant scattering but much smaller than the wavelength of light, so they still can be treated by Rayleigh theory. For this reason, many of the scattering samples used are colloids, mixtures in which small particles of one substance are dispersed in another substance. The dispersed particles are usually between 1 and 1,000 nm, larger than most molecules but too small to be seen, even with a microscope. Table 2-1 lists various types of colloids, classified according to the phases of the dispersed and continuous phases.

The wavelengths of red light emitted by He–Ne and diode lasers fall between 600 and 700 nm, and any colloids with particle sizes significantly smaller than this range would qualify as Rayleigh scatterers. The colloids used in the experiments in this chapter approximate this condition, and Rayleigh theory is used to interpret the data in most of the experiments.

Several experimental difficulties can complicate the interpretation of scattering data. One important difficulty is multiple scattering, in which a photon interacts with several scattering centers before it is detected. Multiple scattering can cause significant deviations from the spatial distribution and polarizations of scattered light predicted by Rayleigh theory. To minimize multiple scattering, the experiments generally use the most dilute solutions possible. Another potential difficulty is that the dispersed particles in colloidal suspensions may interact

TABLE 2-1.   TYPES OF COLLOIDS

Colloid Type	Continuous Phase	Dispersed Phase	Examples
Aerosol (liquid)	Gas	Liquid	Fog, mist, aerosol sprays
Aerosol (solid)	Gas	Solid	Smoke, viruses, airborne bacteria, tear "gas"
Foam	Liquid	Gas	Soap suds, whipped cream, beer foam
Emulsion	Liquid	Liquid	Milk, mayonnaise
Sol or gel	Liquid	Solid	Paint, gelatin, starch, clay, jelly
Solid foam	Solid	Gas	Plastic foams, marshmallow
Solid emulsion or gel	Solid	Liquid	Butter, cheese
Solid sol	Solid	Solid	Ruby glass, some alloys

with each other and form ordered structures in the solution. For instance, experiment 2-4 demonstrates that latex spheres, which are used in several of the scattering experiments, form colloidal crystals under certain conditions. If the dispersed particles have any regular order, then the light scattered would likely demonstrate diffraction effects, which are discussed in the next chapter. Finally, obtaining dust-free solutions free from background scattering is often difficult.

REFERENCES

1. Kerker, M. *The Scattering of Light and Other Electromagnetic Radiation.* Academic; New York: 1969; 31.
2. Wood, R.N. *Physical Optics,* 3rd ed. MacMillan; New York: 1934 (Dover edition, 1967); Chapter XIII.
3. Meyer-Arendt, J.R. *Introduction to Classical and Modern Optics.* Prentice-Hall, Englewood Cliffs, NJ: 1972; Chapter 3.6.
4. Bohren, C.F., Fraser, A.B. *Phys. Teach.* 23 (1985) 267.
5. Braun, C.L., Smirnov, S.N. *J. Chem. Ed.* 70 (1993) 612.
6. Bohren, C.F. *Clouds in a Glass of Beer.* John Wiley & Sons, New York: 1987; Chapter 14.
7. Wood, R.N. *Physical Optics,* 3rd ed. MacMillan; New York: 1934 (Dover edition, 1967); 433.
8. Meyer-Arendt, J.R. *Introduction to Classical and Modern Optics.* Prentice-Hall; Englewood Cliffs, NJ: 1972; 350–351.
9. Shaw, D.J. *Introduction to Colloid and Surface Chemistry,* 3rd ed. Butterworths, London: 1980; 46.

THE TYNDALL EFFECT

The scattering of light by colloidal particles is often intense enough to make a beam of light passing through the medium clearly visible. This phenomenon is known as the Tyndall effect. This experiment uses the Tyndall effect to monitor the growth of sulfur particles resulting from the disproportionation reaction of thiosulfate in acidic solution. In addition, an overhead projector is used as a white light source to demonstrate that the sulfur particles preferentially scatter higher frequency (shorter wavelength) light, as predicted by scattering theory.

DEGREE OF DIFFICULTY

Experimental: easy
Conceptual: easy

MATERIALS

- laser
- 2 volumetric flasks, 1 liter (or other 1-liter containers)
- small test tube
- 1.0 mL of sodium thiosulfate ($Na_2S_2O_3$), 3.0 M
- 1.0 mL of sulfuric acid (H_2SO_4), 3.0 M
- cardboard sheet
- overhead projector
- large Petri dish

PROCEDURE

Demonstration of the Tyndall Effect during Light Scattering

1. Set up the experimental apparatus as shown in Figure 2-1-1.

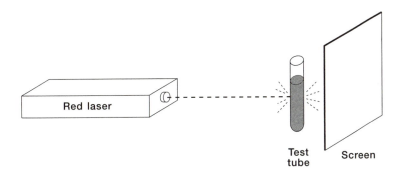

FIGURE 2-1-1. Schematic diagram of experimental setup for observing the Tyndall effect.

25

2. Add 1.0 mL of 3.0-M sulfuric acid to roughly 995 mL of distilled water in a 1-liter volumetric flask. Mix thoroughly.

3. Add 1.0 mL of 3.0 M thiosulfate solution to the volumetric flask, fill to the 1.0-liter mark with distilled water, and mix thoroughly.

4. Transfer 2.0 mL of this solution to a small test tube. Pass the laser light through the solution. As the solution becomes cloudy, note the change in the amount of light scattered at 90° and the defocusing of the transmitted beam over time. Thirty minutes should be enough time to produce high levels of scattering.

Preferential Scattering of Light of Higher Frequency (Shorter Wavelength)

5. Cut a 0.75-in. by 2.5-in. slit in a piece of heavy cardboard and place the cardboard on the surface of an overhead projector so that the light from the projector passes only through the slit.

6. Fill the Petri dish with some of the colloidal solution prepared above, and place the dish over the slit opening.

7. The transmitted projection light should appear yellow–red on the screen. Occasionally, scattered blue light appears on the periphery of the screen.

HAZARDS AND PRECAUTIONS

Observe the proper precautions for handling acids. Wear eye protection at all times.

DISPOSAL

Flush the solution into the sink with flowing water.

DISCUSSION

An everyday example of the Tyndall effect is that observed when you drive through fog and can see the beams of light from your headlights. Although the individual water droplets in fog are far too small to see unaided and would be difficult to observe using a microscope, because of light scattering we know that scattering centers must be present. In this experiment, the Tyndall effect is used to detect the presence of microscopic sulfur particles in solution and monitor the growth of the particles. To a first approximation, increases in scattering intensity can be explained by the growth in particle size.

The chemical reaction used to create the colloidal particles in this experiment was developed by LaMer and Barnes (1) and consists of growing sulfur particles from thiosulfate in acidic solution:

$$S_2O_3^{-2} \xrightarrow{\text{H}^+} S + SO_3^{-2} \qquad (2\text{-}1\text{-}1)$$

This reaction is an example of a disproportionation reaction, one in which the reactant ($S_2O_3^{-2}$) is both oxidized (to SO_3^{-2}) and reduced (to S). The sulfur particles produced by this reaction gradually grow from single atoms to larger aggregates that separate from solution after a day or so.

In the early stages of the reaction, when the sulfur particles remain small enough to stay in solution, the reaction mixture is a colloid that scatters higher frequency light more than it scatters lower frequency light. Shining white light through the solution, as done with the overhead projector, results in the preferential transmission of red light. This effect should intensify over time, because scattering efficiency increases with particle growth. When the sulfur particles eventually become large enough to separate from solution, the scattered light may show a varying spatial pattern.

Another system frequently used to illustrate the Tyndall effect is the reduction of gold from $HAuCl_4$ to produce particles of various sizes and rich colors. Other solutions such as dilute milk, egg albumin, or latex microsphere solutions also can be used in interesting overhead projector demonstrations (2,3).

REFERENCES

1. LaMer, V.K., Barnes, M.D. *J. Colloid Sci.* 1 (1946) 71.
2. Smith, G.C. "Demonstrations to Illustrate Spectroscopic Principles." Presented at the 199th National Meeting of the American Chemical Society, Boston, April 1990; CHED, Paper No. 133.
3. Goldsmith, R.H. *J. Chem. Ed.* 65 (1988) 623.

DEW POINT TEMPERATURE AND CLOUD FORMATION

A laser is used in two related experiments to investigate changes occurring at the dew point and to examine the importance of nucleation for cloud formation.

DEGREE OF DIFFICULTY

Experimental: easy
Conceptual: easy

MATERIALS

Part 1: Detection of the Dew Point Temperature

- laser
- light-detection system
- magnetic stirrer and stirring bar
- 400-mL beaker
- ring stand
- clamps
- ice
- thermometer (down to at least $-10°$ C)
- dew point apparatus (*optional;* S41859, Fisher-EMD, Chicago, IL; see illustration in catalog)

Part 2: The Nature of Particles in Clouds

- laser
- light-detection system
- cloud-forming apparatus (#1730, Sargent-Welch Scientific, Skokie, IL; see illustration in catalog)
- source of smoke, such as incense, punk, or matches

PROCEDURE

Part 1: Detection of the Dew Point Temperature

1. Set up the apparatus as shown in Figure 2-2-1. Clamp the beaker in place on the stirrer. Add 100 mL of pure water and the stirring bar to the beaker.

2. Direct the laser beam tangent to the beaker. Clamp the thermometer in place.

3. Position the light detector to continuously monitor the amount of laser light scattered from the glass surface in a predetermined, fixed direction.

28

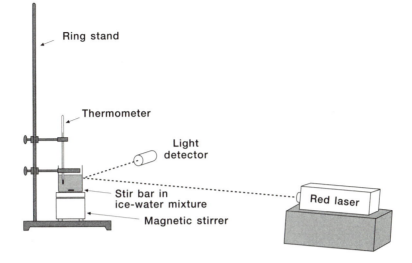

FIGURE 2-2-1. Schematic diagram to show experimental setup for detecting the dew point temperature using laser light scattering from the surface of a beaker.

4. Slowly add ice and monitor changes in the intensity of scattered light as a function of the temperature of the water bath until the intensity of the scattered light changes significantly. Note the point at which condensation is first noticeable visually.

5. Plot the intensity of scattered light vs. the temperature of the bath.

Part 2: The Nature of Particles in Clouds

6. Add 1 mL of water to the cloud-forming apparatus. Pass the beam of the laser through the pear-shaped glass bulb.

7. Align the laser beam with the light detector to measure transmitted light intensity.

8. While holding the apparatus firmly in place, squeeze and relax the rubber bulb. Record the changes in transmitted light intensity.

9. Open the clamp and squeeze the rubber bulb so that some of the air is forced from the apparatus. Place a source of smoke particles near the open end of the rubber tube, and release the bulb so that some of the smoke particles are drawn into the apparatus. Punk or incense may be used to produce the smoke, although an extinguished match may suffice. Once the particles are drawn in, close the clamp on the rubber tube.

10. Repeat step 8. Compare the changes in transmitted light intensity that occur with the smoke particles in the apparatus to those observed before introducing these particles.

HAZARDS AND PRECAUTIONS

Wear eye protection throughout the experiment.

DISPOSAL

All materials used may be washed down the sink.

DISCUSSION

Part 1: Detection of the Dew Point Temperature

The relative humidity of air is the percentage of water vapor contained in the air compared with the maximum amount the air could hold at that temperature (100% humidity). In general, warmer air "holds" more water vapor than cooler air, largely because the vapor pressure of water increases with temperature. As the temperature of a gas that contains water vapor decreases, the relative humidity increases. Eventually, the gas can reach 100% relative humidity, at which point condensation begins. This point is known as the *dew point.* Very humid air reaches the dew point at higher temperatures than dry air. When the air is very dry, the dew point may be less than 0° C, resulting in frost.

In this experiment, a laser is used to detect the dew point by revealing increased scattering caused by the formation of thin water films on the outside of a beaker filled with ice and water. Figure 2-2-2 shows typical experimental results. Notice that increased scattering commences at approximately 3° C, whereas the visual observation of condensation was made at 2° C. The use of a laser allows the detection of subtle effects, such as the formation of thin films, that are not visible to the unaided eye.

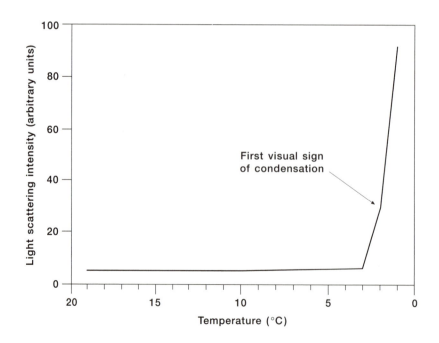

FIGURE 2-2-2. Typical plot resulting from determination of the dew point temperature using laser light scattering as depicted in Figure 2-2-1.

Part 2: The Nature of Particles in Clouds

The cloud-forming apparatus demonstrates the importance of nucleation for cloud formation. Clouds form in the apparatus only when moist air (that may contain smoke particulates) has been introduced into the apparatus, and the pressure is lowered adiabatically. Smoke facilitates the process by providing nucleation for cloud growth. Clouds in the sky do not require smoke to form, but some form of nucleation is necessary for cloud formation.

The use of a laser enhances this demonstration by providing a qualitative measure of the size of the particles in the bulb. Light scattering from the sample increases as the cloud forms, because the average size of the scattering particles grows as water condenses around the smoke particles. This increase in light scattering is detected as a decrease in transmitted light intensity. An interesting He–Ne laser variant of this experiment, also illustrating adiabatic cooling, has been reported in reference 1.

REFERENCE

1. Blaszczak, Z., Gauden, P. *Am. J. Phys.* 58(11) (1990) 1112–1113.

SPATIAL DISTRIBUTION AND POLARIZATION OF SCATTERED LIGHT

A colloidal mixture of the protein albumin in water is used as a scattering medium to study the spatial distribution and polarization of scattered light.

DEGREE OF DIFFICULTY

Experimental: easy
Conceptual: moderate

MATERIALS

- laser
- 2 polarizers
- scattering cell: a test tube, a glass bulb, or a 1-cm^2 cuvette
- 100 mg of albumin (Sigma product #A5378. Approx. $13 for 1 g) (can substitute $\approx$ 100 nm of latex spheres)
- light-detection system (if available)
- glass cylinder and 2 glass plates *(optional)*

PROCEDURE

1. Assemble the apparatus as shown in Figure 2-3-1. The polarizer can be used or withdrawn as needed. The second polarizer is hand held and is not indicated in the figure.

2. Prepare a solution of 100 mg of dry albumin protein in 20 mL of water. From this concentrated solution, prepare several dilute solutions of various concentrations. Experiment with the concentrations used to determine the minimum concentration

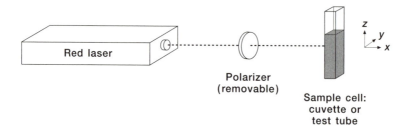

FIGURE 2-3-1. Schematic diagram of experimental setup for observing the spatial distribution and polarization of scattered light.

needed for good scattering with the laser. The use of a very dilute solution minimizes multiple scattering and demonstrates the polarization effects most clearly. This experiment can also be performed with dilute solutions of latex microspheres.

3. With the polarizer removed, direct the light beam from the laser on the scattering cell and view the scattered light in the y and z directions, and at various angles in the xy and xz planes. Determine which regions about the cell have greater and lesser scattered light intensity.

4. Test the scattered light for polarization using the hand-held polarizer by peering through the polarizer along the y axis and rotating the polarizer. Repeat along the z axis.

5. Replace the first polarizer (between the laser and the sample), and test the light intensity in various positions about the scattering center as in step 3. First view the scattered light directly, and then make quantitative measurements at various positions using the light-detection system (if available).

6. Test the scattered light for depolarization by placing the second polarizer along the y or z axis, 90° out of phase with respect to the first polarizer, and view the scattered light passing through the second polarizer. The extent of depolarization may be increased by warming the scattering solution or using a more concentrated solution. Examine the light intensity passing through the second polarizer as a function of the angle of rotation of the second polarizer from the first.

7. *(Optional)* Replace the sample cell with an empty glass cylinder 2 to 6 inches in diameter. Fill the cylinder with smoke (from burning incense or punk, for instance), and cap the end of the cylinder with a glass plate. Observe the scattering caused by the small particles in the smoke. Repeat the observations made in steps 3 through 7.

HAZARDS AND PRECAUTIONS

Wear eye protection throughout the experiment.

DISPOSAL

The albumin and latex spheres solutions may be washed down the sink.

DISCUSSION

Dissolving small amounts of albumin in water creates a colloidal suspension that effectively scatters light. Figure 2-7 in the introduction summarizes the expected results of shining unpolarized laser light through the sample, as in steps 3 and 4. Light is scattered in all directions, but intensity maxima and minima are still observed. In addition, light scattered perpendicularly to the laser beam should be linearly polarized, which can be confirmed by using a hand-held polarizer as described in the procedure. Figure 2-5 depicts the expected scattering patterns from polarized light, as in steps 5 and 6. In this case, light does not scatter in all directions. Instead, regions of zero intensity should be observed in the directions that are both perpendicular to the beam of light and in the plane of the electric field of the incident light. All scattered light should be linearly polarized, which again can be tested with a hand-held polarizer.

In practice, the light scattered from the polarized laser beam will not be perfectly polarized, primarily because of multiple scattering. If photons scatter from more than one center before they leave the cuvette, their polarizations can no longer be predicted in a simple manner. Keeping the concentration of scattering particles low helps minimize multiple scattering. Another experimental difficulty is that laser light from a He–Ne laser is almost never completely unpolarized.

An interesting follow-up experiment is to repeat the experiment using different sizes of latex spheres. As the particle size increases, the experimental observations will likely deviate more and more from the predictions of Rayleigh theory. Even without multiple scattering, light scattered from larger particles is almost never perfectly polarized.

REFERENCES

1. Hecht, E. *Optics,* 2nd ed. Addison-Wesley; Reading, MA: 1987 (reprinted 1990); 278.
2. Meyer-Arendt, J.R. *Introduction to Classical and Modern Optics.* Prentice-Hall; Englewood Cliffs, NJ: 1972; 280.

AGGREGATION OF LATEX MICROSPHERES

Similar to experiment 2-1, this experiment monitors the growth in size of scattering centers by observing the corresponding increase in light scattering. In this experiment, a solution of polystyrene spheres is studied in which the spheres grow into larger and larger clusters upon the addition of a salt solution.

DEGREE OF DIFFICULTY

Experimental: moderate
Conceptual: moderate

MATERIALS

- laser
- light-detection system
- cuvette, 1 cm^2
- cuvette holder
- 1 drop of latex spheres, 107 nm, 2% solution (Duke Scientific, Palo Alto, CA)
- 1.0 mL of saturated NaCl solution
- 0.45-μm filter (if available)

PROCEDURE

1. Prepare a solution of one drop of 2% latex spheres solution in 30 mL of distilled water. Mix and set aside in a closed container. If possible, pass the distilled water through a 0.45-μm filter to remove possible particulates, and rinse the container with some of the filtered water before preparing the latex spheres solution. (Note the experiment can still be performed if the filter is not available.) Try to use water with as few impurities as possible.

Light Scattering At 90°

2. Set up the scattering apparatus for 90° scattering (Figure 2-4-1).

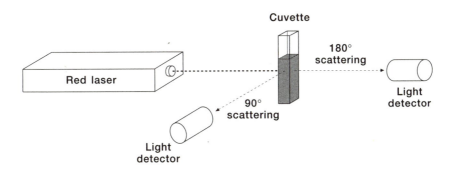

FIGURE 2-4-1. Schematic diagram of experimental setup for observing the aggregation of latex microspheres.

3. Measure the background scattering of the distilled, filtered water. This intensity should be low.

4. Empty the cuvette, add 1.5 mL of the latex sphere solution prepared in step 1, and measure the scattering.

5. Without moving the cuvette, add 0.5 mL of the saturated sodium chloride solution, mix the solution with the pipette, and immediately begin taking periodic intensity measurements. Take readings at least every 20 seconds for the first 5 minutes and every minute for an additional 5 to 8 minutes.

Light Scattering At 180°

6. *(Optional)* Repeat steps 2 through 5 for light intensity at 180°. Take readings every 30 seconds for the first 10 minutes and every minute for an additional 5 to 10 minutes.

7. Plot light intensity vs. time.

HAZARDS AND PRECAUTIONS

Wear eye protection throughout the experiment.

DISPOSAL

The latex spheres solution and NaCl aggregates may be washed down the sink.

DISCUSSION

Under certain conditions, latex spheres in suspension form clusters, as in this experiment, or ordered crystalline structures, which are studied in experiments 3-3 and 3-4. These structures can be explained in terms of attractive and repulsive forces between the latex spheres. Although the polystyrene of which the spheres are composed is electrically neutral, latex spheres usually are manufactured with relatively few ionizable groups on their surfaces. For instance, the latex spheres we used were sulfonated and thus had a negative surface charge. Because all of the spheres are of like charge, they repel each other electrostatically. Counteracting the electrostatic repulsions are attractive forces between the spheres such as dipole–dipole, van der Waals, and London dispersion forces. If the attractive forces dominate the repulsive forces, then the spheres aggregate into clumps. If the repulsive forces dominate, the spheres either remain separate or organize into crystalline structures that minimize the repulsive interactions.

The ionic strength (electrolyte concentration) of the suspension of latex spheres strongly affects the balance between the attractive and repulsive potentials. As shown in Figure 2-4-2, at low ionic strength, the electrostatic repulsions between the spheres dominate the attractive forces for most values of *r*, the intersphere distance. A region of attractive interaction occurs between the spheres (represented as a "potential well"), but a large repulsive barrier prevents the spheres from approaching each other close enough to be attracted to each other. Increasing the ionic strength has little effect on the attractive forces but causes a dramatic decrease in the repulsive potential (1), because the

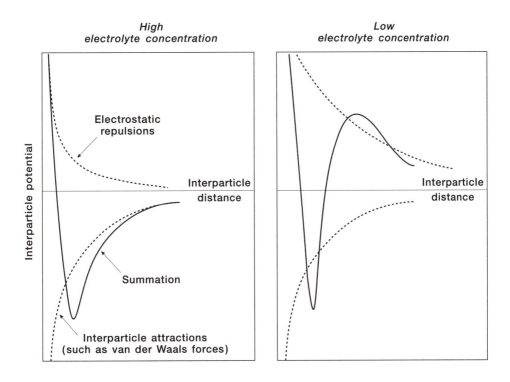

FIGURE 2-4-2. Interactions between latex spheres at high and low electrolyte concentration.

positive and negative ions in solution partially shield the spheres from each other electrostatically. Thus, as the ionic strength increases, the barrier height diminishes and the spheres are free to aggregate.

In this experiment, laser light scattering is used to monitor the aggregation of latex spheres in suspension after the addition of a salt solution. The intensity of light scattered at 90° to the incident laser beam increases over time (Figure 2-4-3), which reflects the increasing number density and size of the aggregates.[1] Figure 2-4-4 shows the changes in the intensity of light at 180°. Keep in mind that the intensity of light at 180° is a combination of transmitted light (light that did not interact with any particles) and light that happened to scatter at 180°. Thus, although light intensity at 180° decreases with time at roughly the same rate that scattering at 90° increases, the light intensity at 180° may never go to zero.

---

1. For several reasons, these data cannot be used to model quantitatively the growth of the aggregates. First, the extent to which the increasing size and increasing number density are responsible for the increased light scattering is impossible to determine. Second, at any time, a wide range of sizes and even shapes of aggregates are likely present in the suspension. In addition, the aggregates are too large to be treated accurately by Rayleigh theory. More sophisticated light scattering methods perhaps could overcome some of these difficulties.

38

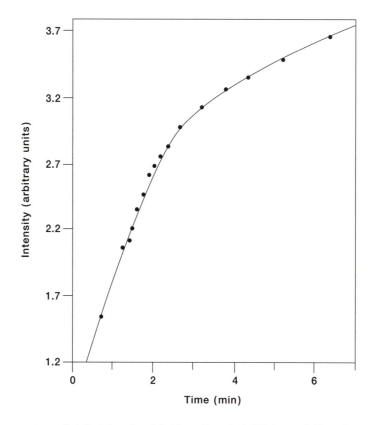

FIGURE 2-4-3. Intensity of light scattered at 90° to a solution of aggregating latex microspheres.

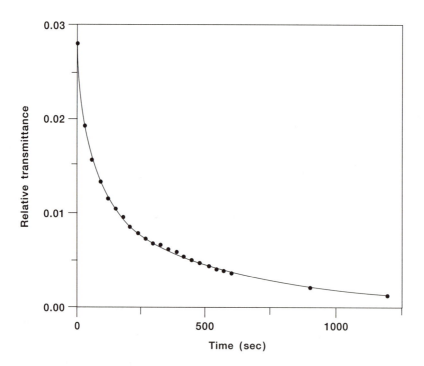

FIGURE 2-4-4. Intensity of light scattered at 180° to a solution of aggregating latex microspheres.

REFERENCE

1. Sunner, J. Nishizawa, K., Kebarle, P. *J. Phys. Chem.* 85 (1981) 1814.

EMULSIONS AND MICROEMULSIONS

A stable, transparent emulsion known as a microemulsion is prepared from a milky white emulsion by adding an agent called a cosurfactant. A laser is used to locate the transition point from a cloudy to a clear suspension.

DEGREE OF DIFFICULTY

Experimental: easy
Conceptual: moderate

MATERIALS

- laser
- burette
- sample cell: glass vial, at least 15-mL capacity, large enough to hold stir bar
- magnetic stirrer and small stir bar
- 2.0-M KOH solution, 1.5 mL
- 0.208 g of oleic acid (Sigma product #O3879. Approx. $7 for 1 g)
        or
- 0.220 g of myristic acid (Sigma product #M3128. Approx. $7 for 10 g)
- n-dodecane, 1.5 mL (Sigma product #D4259. Approx. $6 for 25 mL)
- 1-pentanol, 1 mL (Sigma product #P8274. Approx. $9 for 100 mL)
- HCl, 1.0 M (optional)
- light-detection system (optional)

PROCEDURE

1. Pour 5 mL of dodecane and 5 mL of water into a small beaker. Mix them thoroughly and observe. Adding food coloring to the water may help in visualizing the changes in the two components.

2. Set up an apparatus for titration by clamping the sample cell on a magnetic stirrer base and adding a stir bar to the cell as shown in Figure 2-5-1. Position the laser so that the beam passes through the sample cell.

3. Fill the burette with 1-pentanol.

4. Add 0.208 g oleic acid (or 0.220 g myristic acid) to the sample cell.

5. Add 7.0 mL of distilled water and 1.5 mL of the KOH solution to the cell and begin gentle stirring. Direct the laser beam through the sample cell so that the path of the beam in the cell is free of the stirring vortex.

6. Add 1.5 mL of dodecane (1.0 mL if myristic acid is used) to the titration cell. The resulting solution should be a milky white emulsion.

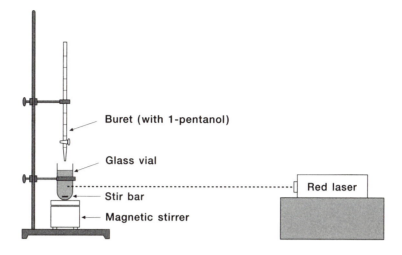

FIGURE 2-5-1. Schematic diagram of experimental setup for preparing a microemulsion from a coarse emulsion and detecting the endpoint with a laser.

7. Reposition the laser beam, if necessary, and begin titration with 1-pentanol, adding the titrant one drop at a time. Find the endpoint of the titration by noting when the light scattering attains a minimum. You can find the endpoint by using a light-detection system positioned for 90° scattering or by observing the rapid focusing of the transmitted beam on a screen. The amount of 1-pentanol added is typically 0.40 to 0.50 mL, or 8 to 10 drops.

8. After observing the relatively clear, nonscattering microemulsion, continue dropwise titration to produce a bluish emulsion, and observe the increased light scattering intensity.

9. *(Optional)* Break the emulsion in the titration cell by adding a few drops of 1.0-M HCl. This procedure converts the potassium salt of the acid to the protonated acid, which has reduced surfactant properties.

HAZARDS AND PRECAUTIONS

The KOH solution is slippery to the touch. Flush the skin quickly with large amounts of water if exposed to the KOH. Pentanol and dodecane are not highly toxic, but avoid breathing their vapors. Oleic and myristic acid are skin, lung, and eye irritants. Wear eye protection throughout the experiment.

DISPOSAL

Place the reaction mixtures in an organic-waste container, and dispose of them properly.

DISCUSSION

Mixing together polar (hydrophilic) and nonpolar (hydrophobic) liquids generally creates a heterogeneous mixture that has components that are clearly separated into two

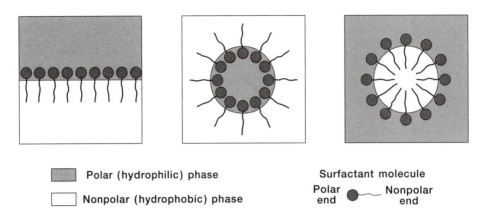

FIGURE 2-5-2. Surfactant molecules aligned at the interface between polar and nonpolar phases.

phases, such as oil droplets in water. This type of mixture is created in step 1 of the procedure with dodecane and water. Because water is highly polar and *n*-dodecane is nonpolar, the two liquids do not mix together well. Even after thorough mixing, the two liquids always separate into distinct phases.

The addition of compounds known as soaps or surfactants, however, can stabilize globules in these heterogeneous solutions. A surfactant, or surface-active agent, accumulates at the interface between the polar and the nonpolar phases and acts to reduce the surface tension. Surfactants stabilize polar–nonpolar mixtures because they are *amphiphilic,* meaning that they have polar-attracting and nonpolar-attracting regions. The hydrocarbon chains of the surfactant typically align perpendicularly to the interface, as shown in Figure 2-5-2; the polar end of the molecule aligns in the hydrophilic phase and the nonpolar end aligns in the hydrophobic phase. The oleic and myristic acids used in this experiment are surfactants, and their structures are shown in Figure 2-5-3.

The term *emulsion* refers to a liquid–liquid mixture in which the liquid "drops" are of colloidal dimensions. Surfactants are known as emulsifiers in such systems because they enhance the formation of emulsions from two distinct phases. Homogenized milk is a common emulsion, as is mayonnaise, in which the egg yolk lipids coat the oil droplets as a monolayer. The emulsion in this experiment is essentially a water–dodecane emulsion, and the potassium salt of oleic or myristic acid acts as the emulsifier.

Most emulsions separate into two phases over time, but under certain conditions, stable emulsions known as *microemulsions* can be prepared that do not show a tendency to separate (1). As the name suggests, microemulsions have dispersed phases composed of very small particles. The use of dynamic light scattering has allowed researchers to estimate particle size for microemulsions to be approximately 5 nm (2,3).

A stable, transparent microemulsion is prepared in this experiment by adding 1-pentanol to the original milky white emulsion, which is called a coarse or hard emulsion. 1-Pentanol acts as a cosurfactant in this experiment because it works with the surfactant to stabilize the emulsion. The cosurfactant plays a complex role in the formation of the microemulsion (4). Similar to the surfactant, the cosurfactant (1-pentanol) has a polar end and a nonpolar end, and for this reason it accumulates at the interface

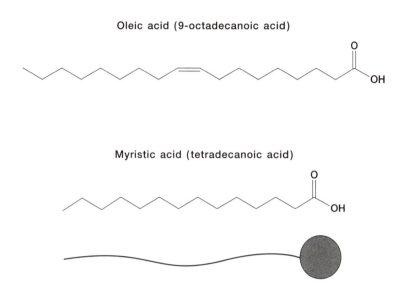

FIGURE 2-5-3. Molecular structures of the surfactants oleic acid and myristic acid.

between the polar and nonpolar phases along with the surfactant. Because the cosurfactant is a much smaller molecule than the surfactant, cosurfactant molecules can fit between the surfactant molecules (Figure 2-5-4). Thus, the cosurfactant helps lower the
surface tension between the two phases more than the surfactant does alone and helps
promote the formation of smaller droplets of the nonpolar phase (n-dodecane).

Coarse emulsions strongly scatter light, but microemulsions scatter light only very
weakly, because of their small particle sizes. As the cosurfactant 1-pentanol is added,

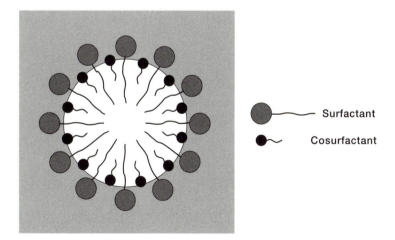

FIGURE 2-5-4. Cosurfactant molecules accumulated among surfactant molecules at the interface between polar and nonpolar phases.

light scattering gradually decreases and reaches a minimum at the transition to a microe-mulsion. Thus, laser light scattering can be used to forecast and detect the transition between the coarse emulsion and the microemulsion. The addition of excess 1-pentanol results in formation of a new emulsion that appears blue and contains visible droplets that are probably excess 1-pentanol that has separated from the emulsion. This experi-ment can be expanded in scope by examining the effects of using different concentra-tions of acid or different alcohols and oils as cosurfactants.

REFERENCES

1. Rosano, H.L., Lan, T., Weiss, A. *J. Colloid Int. Sci.* 72 (1979) 233.
2. Cazabat, A.M., Langevin, D., Pouchelon, A. *J. Colloid Int. Sci.* 73 (1980) 1.
3. Bellocq, A.M., Fourche, G. *Optica Acta* 27 (1980) 1629.
4. Myers, D. *Surfactant Science and Technology.* VCH Publishers; New York: (1988); 176.

DYNAMICS OF THE GEL-FORMATION PROCESS

Light scattering at 90° from an incident laser beam is used to monitor a gel-formation reaction.

DEGREE OF DIFFICULTY

Experimental: moderate
Conceptual: moderate

MATERIALS

- laser
- light-detection system
- beam splitter
- hot plate or portable hairdryer
- cuvette, 1 cm^2
- cuvette holder
- 0.45-μm and 0.22-μm filters
- thermometer
- gelatin
- 0.15-M NaCl solution

PROCEDURE

1. Set up the apparatus as shown in Figure 2-6-1.

2. Filter the 0.15-M NaCl solution through a 0.22-μm filter. Prepare 100 mL of gelatin solution using 3.0 g gelatin and 0.15-M NaCl solution. Warm gently at 55° C for 1 hour. Filter 10 mL of this warmed solution, while still in the liquid form, through a 0.45-μm filter.

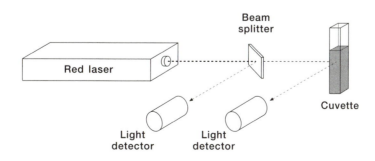

FIGURE 2-6-1. Schematic diagram of experimental setup for monitoring the formation of gelatin.

3. Add the liquid gelatin to the cuvette and place the cuvette in the cuvette holder at room temperature. Align the beam of the laser to pass through the gelatin.

4. Record the intensity of the scattered light every 5 minutes for up to 3 hours, or use a chart recorder to record intensity continuously. Although lowering the temperature below room temperature shortens the gelation time, maintaining a constant temperature may be difficult.

5. Warm the sample cell using a portable hairdryer or hot plate; be careful not to heat the gelatin too quickly. Do not stir the melting gelatin. Record the intensity of the scattered light every minute until the gelatin is melted and the intensity reaches a new steady value. Record the temperature of the solution.

6. *(Optional)* Repeat steps 1–5 using different gelatin concentrations.

HAZARDS AND PRECAUTIONS

Wear eye protection throughout the experiment.

DISPOSAL

All materials may be washed down the sink.

DISCUSSION

The "gelatin" sold in supermarkets is prepared from the protein collagen. Proteins are large organic molecules composed of units called amino acids linked together in long chains (Figure 2-6-2). Approximately 20 amino acids are commonly found in living organisms, and collagen molecules are composed of a dozen or more different amino acids. The amino acids glycine, proline, and hydroxyproline (Figure 2-6-3) are the most abundant in collagen (1); together they account for almost two-thirds of the amino acids in the protein (see Table 2-7-1 for the complete composition of gelatin).

Amino acid structure

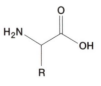

FIGURE 2-6-2. Amino acid and protein structure. The *R* group can be one of more than 20 side groups.

Protein structure

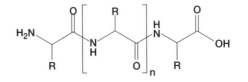

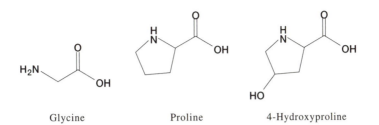

Glycine   Proline   4-Hydroxyproline

FIGURE 2-6-3. Molecular structures of the three most common amino acids in gelatin: glycine, proline, and hydroxyproline.

The general structure of collagen is depicted in Figure 2-6-4. Three individual strands of protein wrap around each other to form a triple helix, although the ends of collagen often consist of nonhelical regions. The three protein strands are held together primarily by hydrogen bonding of the —OH group on hydroxyproline. Each strand consists of more than 1,000 amino acids, and the length of the molecule is typically 3,000 Å, whereas the diameter is approximately 15 Å (1). In nature, collagen molecules wrap together to form strong fibers, which are among the most important structural materials in animals. In fact, collagen is the most abundant protein in mammals, constituting approximately one-fourth of their weight (1).

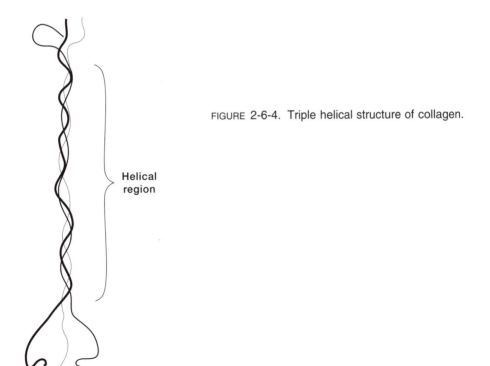

Helical region

FIGURE 2-6-4. Triple helical structure of collagen.

When proteins are heated, their three-dimensional structure is often irreversibly altered. This process is known as *denaturation*. When collagen molecules are heated in aqueous solution, the three individual strands of protein dissociate from each other and the triple helical structure is lost. The gelatin packets bought in grocery stores contain denatured collagen, which is prepared by boiling the skin, bones, and tendons of animals with water. Gelatin can be dissolved in hot water, and when the gelatin solution cools, the mixture may appear to solidify if the gelatin concentration is high enough. The solidified gelatin is not a true solid but belongs to a class of solid-in-liquid colloids known as *gels*. Gels differ from *sols* in that gel particles are linked in a matrix of some strength, whereas in a sol, the solid particles are randomly dispersed. The intermolecular forces that link gel particles together are a complex combination of electrostatic, dipole–dipole, hydrogen-bonding, and London-dispersion interactions (2). The solid phase in a gel does not crystallize, however, because dissimilar units along the chains prevent much structural regularity.

In gelatin, the denatured, independent protein strands form interstrand helices that give rise to the structural strength of the gel. The original triple helices are not reformed, however (3). The process of gel formation is very complex; it consists of several phases, each of which has different reaction kinetics (3,4). In this experiment, no attempt is made to distinguish between the various stages of gel formation. Instead, the formation of gelatin is monitored qualitatively, using light scattering from a low-power laser. As the colloidal scattering centers grow during gelation, the intensity of scattered light rises steadily to a final plateau, as plotted in Figure 2-6-5.

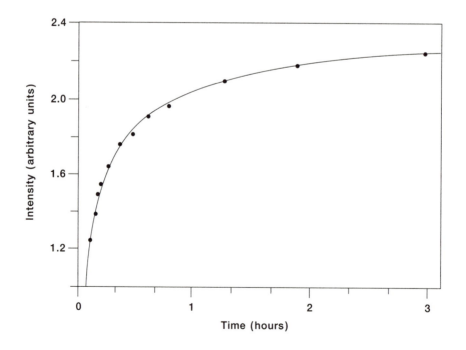

FIGURE 2-6-5. Intensity of light scattered at 90° from a cooling gelatin solution.

Gelatin belongs to a class of gels known as *thermoreversible* gels, gels that undergo reversible gel-to-liquid transitions at a specific temperature. The "melting" of the gel form, which is studied in step 5 of the procedure, corresponds to the breakdown of the intermolecular forces that give the gel its structural strength. During the melting process, scattering intensity may increase initally before it decreases significantly as the gel becomes a liquid. As noted by Boedtker and Doty (5), the initial increases in the scattering intensity may be caused by an increase in disorder at the outset of melting.

Not surprisingly, scattering intensity depends on the concentration of the gelatin solution. Boedtker and Doty (5) report that scattering decreases with increasing gelatin concentration for the gel form and that scattering is directly proportional to gelatin concentration for the melted form. Consequently, the change in scattering intensity that accompanies the melting of the gel decreases with increasing gelatin concentration.

Gels can be formed from a variety of macromolecules, such as sulfated carrageenans, agarose, and pectins. Observing how the gelation of these substances differs from that of gelatin might be interesting. In addition, please refer to experiment 2-7 for further study of gelatin.

REFERENCES

1. Stryer, L. *Biochemistry,* 3rd ed. W.H. Freeman; New York, 1988; 261–274.
2. Burchard, W. *Br. Polymer J.* 17 (1985) 154.
3. Djabourov, M., Maquet, J., Theveneau, H., Leblond, J., Papon, P. *Br. Polymer J.* 17 (1985) 169.
4. Croome, R.J. *J. Photo. Sci.* 22 (1974) 239.
5. Boedtker, H., Doty, P. *J. Phys. Chem.* 58 (1954) 968.

THE ISOELECTRIC POINT OF GELATIN

The electrostatic forces between the collagen strands in gelatin vary with pH, and laser light scattering can be used to determine the pH at which electrostatic interactions are at a maximum. The point at which this maximum occurs is known as the *isoelectric point.*

DEGREE OF DIFFICULTY

Experimental: moderate
Conceptual: moderate

MATERIALS

- laser
- light-detection system
- cuvette holder
- beam splitter
- Erlenmeyer flask, 100 mL
- disposable cuvette, 1 cm^2
- pipettes
- pH meter or indicator paper
- gelatin
- temperature bath
- 1-M HCl, 1 mL
- 1-M NaOH, 1 mL
- 0.45-μm filter

PROCEDURE

1. Filter 125 mL of water through a 0.45-μm (or 0.22-μm) filter and warm it to approximately 55° C. Prepare 100 mL of a gelatin solution that is 0.14% gelatin by weight using this filtered, distilled water. Stir the gelatin solution until well dissolved, then filter it with the 0.45-μm filter. Adjust the temperature to 25° C by placing the flask in a temperature bath.

2. Adjust the pH of the gelatin solution to 9 by adding NaOH, one drop at a time.

3. Set up the experimental apparatus as in experiment 2-6 (Figure 2-6-1). Transfer some of the gelatin solution to the cuvette, and align the laser beam through the gelatin.

4. Determine the intensity of the scattered light at 90° for the sample at pH 9.

5. Pipette the sample out of the cuvette and rinse the cuvette without moving it.

6. Add HCl one drop at a time to the remaining gelatin solution to adjust the pH downward by a unit or so. Stir well. Repeat steps 3–5. Continue this process until the pH

of the gelatin solution is between 3 and 4. Be sure there are no visible signs of gel formation.

7. Plot the scattered light intensity at 90° vs. pH.

HAZARDS AND PRECAUTIONS

Wear eye protection at all times. Wash any HCl or NaOH spills or splashes immediately with water.

DISPOSAL

All materials may be washed down the sink.

DISCUSSION

Please refer to experiment 2-6 for background information on gelatin and proteins.

Amino acids exist as ions in aqueous solution at any pH. Consider, for instance, a simple amino acid such as alanine, which has only two ionizable groups: the carboxylic acid group (—COOH), which can be deprotonated to become —COO⁻, and the amine group (—NH$_2$), which can be protonated to become —NH$_3^+$ (Figure 2-7-1). The carboxylic acid group on alanine has a pK_a^1 value of approximately 2 and the protonated

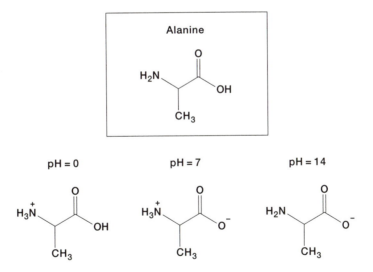

FIGURE 2-7-1. Molecular structure of alanine and the forms of alanine that predominate in aqueous solution at pH 0, 7, and 12.

---

1. The equilibrium of an acid in water can be expressed in terms of an equilibrium constant, K_a, also called the acidity constant, where $K_a = \dfrac{[H_3O][Base]}{[Acid]}$. The p$K_a$ value equals the negative of the logarithm of K_a to the base 10, that is, p$K_a = -\log K_a$.

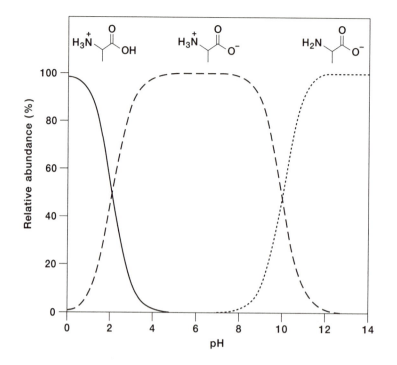

FIGURE 2-7-2. Abundances of the three forms of alanine (shown at top) in aqueous solution as a function of pH.

amine group has a pK_a value of approximately 10. At very low pH (e.g., pH = 0), nearly all of the carboxylic acid and amine groups are protonated, which means that nearly all of the amino acid molecules are cations. By contrast, at high pH (e.g., pH = 12), most of the molecules of amino acid exist as anions, because most of the carboxylic acid and amine groups are deprotonated. At intermediate pH (near pH = 7), most of the molecules of amino acid exist in zwitterionic form, which means that the carboxylic acid group is deprotonated but the amine group is protonated (in effect, the proton from the carboxylic acid group protonates the amine group). The zwitterion has no net charge because it has one positive charge and one negative charge. Figure 2-7-2 shows a graph of the relative concentrations of the three different forms of alanine over a pH range from 0 to 14.

The *isoelectric point* of an amino acid is the pH at which the net charge on all of the molecules of that amino acid is zero. For alanine, the isoelectric point corresponds closely with the pH at which the concentration of the zwitterionic form is at a maximum. The isoelectric point of alanine is 6.0, roughly an average of the pK_a values of the two ionizable groups. Alanine belongs to the class of neutral amino acids, which signifies that the side chain contains no groups that can be protonated or deprotonated in aqueous solution (no ionizable groups). All of the neutral amino acids have isoelectric points near 6. Five common amino acids have ionizable side groups (Figure 2-7-3), and the presence of these side groups affects the isoelectric point. The two amino acids with

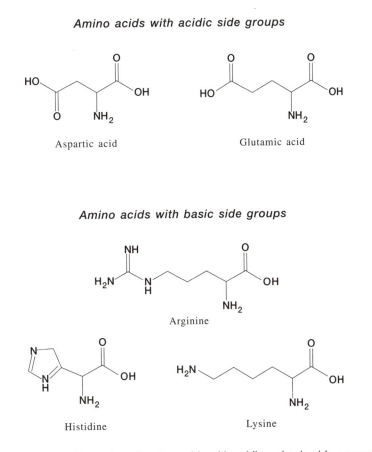

FIGURE 2-7-3. Examples of amino acids with acidic or basic side groups.

acidic side groups have isoelectric points significantly below 6, and the three amino acids with basic side groups have isoelectric points above 6.

The gelatin used in this experiment consists of strands of protein comprising approximately 1,000 amino acids each. Each protein strand has many ionizable groups, which include the carboxylic acid and amine groups on opposite ends of the protein (see Figure 2-6-2) and the side groups on the amino acids. At the isoelectric point of a gelatin solution, the net charge on most of the denatured collagen strands is nearly zero, which minimizes the electrostatic repulsions between the strands. Although the solution is too dilute to allow for aggregation, at the isoelectric point the collagen strands have greater attraction for each other and spend more time in the immediate vicinity of other strands (1). The intensity of light scattering increases as the isoelectric point is approached because the effective size of the collagen particles is larger.

A plot of intensity vs. pH yields the pH at the isoelectric point, which should be approximately pH 5.0 to 5.1. The fact that this isoelectric point is below 6 reflects in part that the number of amino acids with acidic side groups in collagen outnumbers those with basic side groups, as shown in Table 2-7-1.

TABLE 2-7-1.   AMINO ACID CONTENT OF GELATIN[a]

Amino Acid	Approximate Abundance in Gelatin (%)	Comments
Glycine	25.5	
Proline	18.0	
Hydroxyproline	14.1	
Glutamic acid	11.4	Acidic side group
Alanine	8.7	
Arginine	8.1	Basic side group
Aspartic acid	6.6	Acidic side group
Lysine	4.1	Basic side group
Leucine	3.2	
Valine	2.5	
Phenylalanine	2.2	
Threonine	1.9	
Isoleucine	1.4	
Methionine	1.0	
Histidine	0.8	Basic side group

[a] From reference 2, with permission.

REFERENCES

1. Boedtker, H., Doty, P. *J. Phys. Chem.* 58 (1954) 968.
2. *The Merck Index: An Encyclopedia of Chemicals, Drugs, and Biologicals,* 10th ed. Merck and Co.; Rahway, NJ: 1983; 4243.

EXPERIMENT **2-8**

SCATTERING OF LIGHT BY OPTICALLY ACTIVE COLLOIDS

Polarized laser light directed through an optically active colloid produces a spiraling pattern of scattered light, which demonstrates the effect of optical rotation. This experiment is especially well suited for use as a classroom demonstration.

DEGREE OF DIFFICULTY

Experimental: moderate
Conceptual: moderate

MATERIALS

- laser
- safety bucket
- epoxy
- microscope slide
- glass tube, 1-in. diameter by 2-ft. length, with stopper
- adjustable mirror
- ring stand
- large weight (to stabilize ring stand)
- polarizer
- funnel
- filter paper
- 20 g of quinine sulfate (Sigma Product #Q1250. Approx. $25 for 25 g) and 100 mL of glacial acetic acid
        or
- 100 g of sucrose

PROCEDURE

1. Seal a microscope slide to one end of the glass tube with epoxy, making sure the seal is complete and that laser light can pass through the microscope slide. Allow the bond to harden overnight. Use a newly sealed slide each time this experiment is performed. Never leave the quinine sulfate solution in the tube for more than one hour, because the epoxy bond will weaken significantly and at some point, the bond could fail. This problem is greatly reduced when the sugar solution is used, but use the safety bucket with either solution.

2. In the fume hood, dissolve 20 g of quinine sulfate in 100 mL of concentrated acetic acid. Gravity filter the solution. Using a funnel, very slowly pour the solution into the glass tube. Seal the open end of the tube with a rubber stopper. This experiment

55

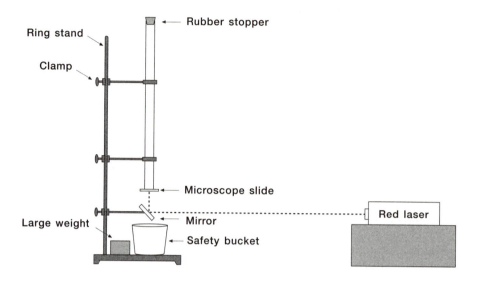

FIGURE 2-8-1. Schematic diagram of experimental setup for observing the scattering of light from optically active colloids.

can also be performed using a highly concentrated solution of sucrose in water. The high concentration of sucrose is required because it has a lower specific rotation than quinine. Performing the experiment with sucrose generally results in less well defined scattering patterns than those when quinine sulfate is used. In other words, the sucrose solutions provide a less stunning demonstration of optical activity.

3. Set up the apparatus shown in Figure 2-8-1. Clamp the tube vertically (stoppered end up) to a ring stand using at least two clamps. Place a large weight on the base of the ring stand for stability and place the safety bucket under the tube. Use the mirror to deflect the laser beam upward along the central axis of the tube. Turn off the room lights, and check for proper alignment of the beam, adjusting the mirror so that the laser beam passes straight through the tube.

4. Observe the scattering from the colloid along the vertical axis of the tube.

5. Place the polarizer in the path of the laser beam as shown in Figure 2-8-2, and observe the scattering from the colloid. Slowly rotate the polarizer and observe the movement of the helix of scattered light.

6. If the quinine sulfate–acetic acid system was used, return the tube to the fume hood when finished with the experiment. Using a funnel, carefully pour the solution into a flask or bottle for storage. Remove the slide and rinse the tube out thoroughly.

HAZARDS AND PRECAUTIONS

Glacial acetic acid causes severe burns upon exposure. Use a safety bucket under the tube at all times. The epoxy securing the slide on the end of the tube is a potential weakness that must be watched carefully. Always remove the slide after each use and

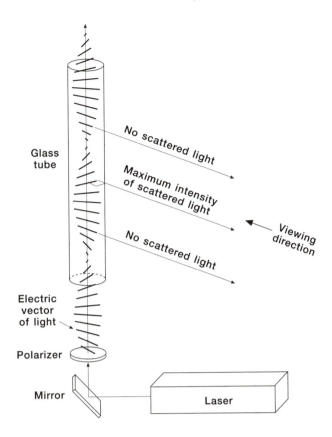

FIGURE 2-8-2. Rotation of the electric field vector of a polarized beam of light through an optically active colloid.

reseal a clean slide to the tube. Quinine sulfate is a skin irritant. Wear safety glasses and gloves at all times.

DISPOSAL

Dilute the glacial acetic acid solutions with large amounts of water and dispose of the very dilute solution in the laboratory sink drain. Sucrose solutions may also be disposed of down the laboratory sink drain, using running tap water.

DISCUSSION

Two compounds are called *isomers* if they have the same molecular formula but different chemical structures. *Optical isomers* are those in which two compounds have not only the same molecular formula but also identical bonding connections between the various atoms. A pair of optical isomers remain distinct from each other, however, because they are nonsuperimposable mirror images of each other. One optical isomer cannot be superimposed on the other, just as your left hand cannot be superimposed on your right hand. Compounds that exist as optical isomers are frequently referred to as *chiral* compounds, and each member of a pair of optical isomers is named an *enantio-*

Mirror plane

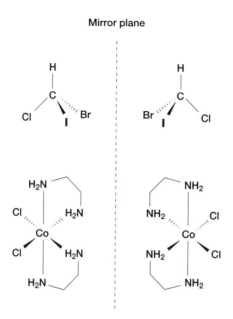

FIGURE 2-8-3. Examples of chiral compounds.

mer. Molecules such as H_2O and CH_4, which do not exist as nonsuperimposable pairs, are called *achiral.* A few examples of chiral compounds are presented in Figure 2-8-3.

Enantiomers have identical atoms and bonds, but the two different forms have different optical properties. If plane polarized light is passed through a solution of a chiral compound, the plane of polarization of the light is rotated either clockwise or counterclockwise. The extent of this rotation depends on the nature of the compound and the path length of the solution, as well as environmental factors such as temperature. Under the same conditions, enantiomers rotate light to the same extent, except that one rotates the plane of polarized light to the left and the other structure rotates it to the right.

Only chiral compounds rotate light in this way, and for this reason chiral compounds are often called *optically active* compounds (and achiral compounds are referred to as *optically inactive*). The reason chiral compounds rotate polarized light is complex and is fully explained only by a quantum mechanical treatment of the interaction of electromagnetic radiation with chiral compounds. One way of thinking about optical rotation, however, is to consider linearly polarized light as a superposition of right-handed and left-handed circularly polarized light. Because circularly polarized light has a handedness like chiral molecules, it is not surprising that right- and left-handed circularly polarized light interact differently with chiral molecules. Specifically, the index of refraction of a chiral substance is different for left- and right-handed circularly polarized light. Thus, the right- and left-handed components of linearly polarized light travel through a chiral medium with different velocities, and one handedness of light is retarded with respect to the other. The net effect of this retardation is to rotate the direction of the polarization of the linearly polarized light.

No easy method can predict whether a particular chiral compound will rotate light clockwise or counterclockwise. Distinguishing between two enantiomers by the direction in which they rotate the polarization of light is often convenient. By convention, left-rotating enantiomers are labeled $(-)$ and right-rotating ones $(+)$. Some chiral compounds rotate light more than others, and chiral compounds can be assigned specific rotation values (often designated in tables by the symbol $[\alpha]$), which tell to what extent the compound rotates light. The specific rotation is defined as

$$[\alpha] = \frac{\alpha}{\ell c} \qquad \qquad (2\text{-}8\text{-}1)$$

where α refers to the rotation (in degrees) the electric field vector of the light undergoes in traveling a distance ℓ (in dm) through a solution with a concentration c (in g/mL).

The specific rotation of a solution should be independent of the concentration of the solution, because implicit in the definition of specific rotation is the assumption that the rotation of the light is directly proportional to the concentration of the solution. The specific rotation of a solution does depend, however, on the temperature of the solution and the wavelength of light traveling through it. Thus, specific rotations are often labeled with a superscript that indicates the temperature (in °C) and a subscript that indicates the wavelength of light (in nm). The specific rotation of the solutions in this experiment might be reported as

$$[\alpha]^{25}_{632.8} \qquad \qquad (2\text{-}8\text{-}2)$$

if a He–Ne laser is used, because the wavelength of light produced by a red He–Ne laser is 632.8 nm. In practice, using the results of this experiment to determine values of the specific rotation for quinine sulfate or sucrose would be problematic. The concentrations of these compounds in the suspensions observed is uncertain, because the quinine and sucrose molecules clump together to form colloidal particles.

Quinine sulfate, when dissolved in concentrated acetic acid, produces an optically active colloidal solution. The colloidal suspension is comprised of small particles of quinine surrounded by a highly concentrated solution of quinine sulfate. When polarized light passes through a long tube of this solution, the direction of polarization of the light rotates along the length of the tube, and the electric field of the light spirals up the barrel of the glass tube like a piece of ribbon candy (as seen in Figure 2-8-2). The quinine sulfate solution has a $(-)$ specific rotation and thus we expect a left-handed spiral up the tube.

As explained in the introduction to this chapter, scattered light intensity is generally highest perpendicular to the dipole axis of scattering particles, which in turn should be aligned with the electric field vector of the light. As a result, the light scattered from the solution displays light and dark regions along the length of the tube. When the polarizer in the path of the laser beam is rotated, the light and dark bands move up or down the tube, depending on which way the polarizer is rotated. This band movement corresponds to a rotation of the electrical field spiral shown in Figure 2-8-2. No light and dark regions are observed with unpolarized light, because unpolarized light is equivalent to a superposition of all possible polarizations, whose effects cancel each other.

Further experiments with quinine sulfate could include varying the concentration, pH, temperature, and wavelength of light, and observing the effects of these variables on the specific rotation. The specific rotation of quinine sulfate can also be compared with that of sucrose, which is smaller under similar conditions.

Chirality in Organic and Biological Chemistry

Chirality is an especially important concept in organic chemistry, because a vast number of organic compounds are chiral. The chiral nature of organic compounds usually results from an asymmetric carbon atom, which is one that is bonded to four different substituents. A simple example of a chiral organic molecule is CHClBrI (shown in Figure 2-8-3). The two enantiomers can be thought of as having two of the attached atoms reversed, causing the two forms to be nonsuperimposable. Many organic compounds have more than one asymmetric carbon, each of which is called a *chiral center*. In these compounds, each chiral center has 2, or 2^n total possible configurations, where *n* is the number of chiral centers. Some of these configurations are mirror images of each other and are called *enantiomeric pairs*. Any given pair of these molecules are not necessarily mirror images, however, and in general are called *diasteriomers* (Figure 2-8-4).

The vast majority of important biological compounds, such as sucrose, have at least one chiral center. The structures of sucrose and quinine are shown in Figure 2-8-5. Amino acids and sugars have a special labeling system to distinguish between different diasteriomers. The letters D and L, from the Latin *dexter* (right) and *laevus* (left), are used to indicate how the —H and —OH groups are attached to a particular carbon atom. Despite their names, the labels do not indicate which way the light is rotated.

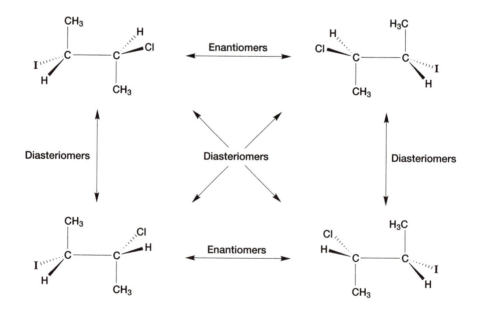

FIGURE 2-8-4. Enantiomeric and diastereomeric relationships among the four isomers of 2-chloro,3-iodobutane.

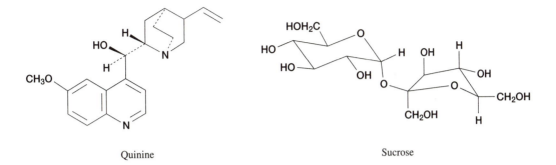

Quinine   Sucrose

FIGURE 2-8-5. Molecular structures of the optically active organic compounds quinine and sucrose.

Interestingly, in almost all life forms, only L forms of amino acids are produced, whereas the D forms of sugars predominate. Although there is no intuitive reason for one configuration to predominate over the other, biological systems have evolved such that for many chiral compounds, only one isomer is observed in nature. This fact has profound implications for the pharmaceutical industry. Because many drugs are chiral, different isomers of the same drug can have entirely different effects on the body (1). One isomer of a drug may have the desired healing effect, while another may have no effect or even be harmful. Thus, controlling the chirality of biological compounds when they are being synthesized is almost always necessary, and chiral selectivity presents a key challenge for the pharmaceutical industry in developing safe and effective drugs.

REFERENCE

1. Borman, S. *Chem. & Engr. News* 68(28) (1990) 9.

RATE OF VESICLE AGGREGATION

Lipids can form colloidal structures in water known as vesicles, which can be considered crude models of biological membranes. In this experiment, you will use laser light scattering to examine vesicle aggregation in a system of dimyristoyl phosphatidylcholine in salt solution at various temperatures.

DEGREE OF DIFFICULTY

Experimental: difficult
Conceptual: moderate

MATERIALS

- laser
- light-detection system
- beam splitter
- cuvette, 1 cm^2
- cuvette holder
- sonicator
- 100-mL beaker
- test tube
- temperature regulator with circulator (optimum), or heated or cooled water bath
- pH meter
- 25 mg of dimyristoyl phosphatidylcholine (Sigma product #P6392. Approx. $11 for 25 mg)
- 1.5 mL of salt solution (NaCl, KCl, Na_2HPO_4, NaH_2PO_4, $MgSO_4$, and glucose)

PROCEDURE

1. Assemble the apparatus as shown in Figure 2-9-1. Because this experiment is lengthy, use a beam splitter to monitor laser power, and take data only when the laser power is at some standard value.

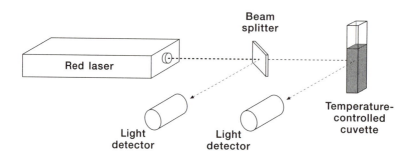

FIGURE 2-9-1. Schematic diagram of experimental setup for observing vesicle aggregation.

62

2. Prepare 100 mL of the salt solution by dissolving all of the following components in enough distilled water to give 100 mL of solution:

Component	Amount (mg)	Concentration in solution (mM)
Sodium chloride (NaCl)	806.5	138.0
Potassium chloride (KCl)	37.3	5.00
Dibasic phosphate (Na_2HPO_4)	86.6	6.10
Monobasic phosphate (NaH_2PO_4)	16.8	1.40
Magnesium sulfate ($MgSO_4$)	12.3	1.02
Glucose	90.1	5.00

Using the pH meter, adjust the pH to 7.4 by adding phosphoric acid. Adjust the temperature of the solution to 30° C with a temperature bath.

3. Add 1.5 mL of the salt solution prepared in step 2 to the 25 mg of lipid (dimyristoyl phosphatidylcholine) in a test tube. This addition will produce a solution that is 24 mM with respect to the lipid.

4. Clamp the test tube in the sonicator and sonicate the solution for 1 hour or until the solution becomes clear. To reduce time, try to find a "hot spot" in the sonicator where agitation is at a maximum.

5. Adjust the solution to room temperature, fill the cuvette, and begin taking light-intensity measurements. Continue taking measurements periodically for approximately 1 hour.

6. (Optional) Repeat the kinetic measurements in step 5 at different temperatures. Use a temperature bath to maintain the temperature of the solution as constant as possible during each run. Keep in mind that best results are obtained if the cuvette is not moved during the course of the measurements. Use several temperatures within 15° of the initial temperature.

HAZARDS AND PRECAUTIONS

Wear eye protection throughout the experiment.

DISPOSAL

All solutions can be washed down the sink with running water.

DISCUSSION

Lipids are organic molecules that consist of a nonpolar alkyl end and a polar or ionic head group. The structure of dimyristoyl phosphatidylcholine is shown in Figure 2-9-2. The head group is electrically neutral, and the head groups and two alkyl chains are connected to a glycerol unit.

Lipids are driven into aggregation by attractive forces among the nonpolar ends and between the ionic head groups and the aqueous environment. Lipids aggregate in several

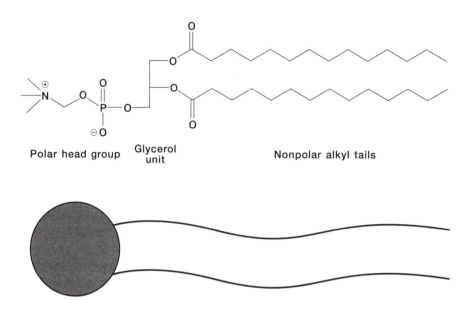

Polar head group   Glycerol
                       unit   Nonpolar alkyl tails

FIGURE 2-9-2. Molecular structure of dimyristoyl phosphatidylcholine, depicting the polar head group and two nonpolar tails.

forms. One is the *micelle,* shown in Figure 2-9-3, which is the common form for soaps and detergents. The shape of the micelle requires the nonpolar group of the lipid to be compact. Lipids that have larger nonpolar groups aggregate into either monolayers or bilayers, which are depicted in Figure 2-9-4. *Vesicles* consist of a bilayer wall that separates an interior environment from the aqueous exterior. Some vesicles closely resemble cell membranes, which are actually bilayer lipids that contain proteins, carbohydrates, or other macromolecules.

In this experiment, the lipids in the original buffered solution of dimyristoyl phosphatidylcholine initially aggregate into monolayers, which are then converted to bilayer

FIGURE 2-9-3. Structure of a micelle, one form in which lipids aggregate.

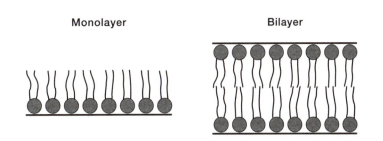

Monolayer   Bilayer

FIGURE 2-9-4. Structures of a monolayer and a bilayer formed by lipid aggregation.

vesicles by sonication. The vesicles produced have diameters of approximately 20 nm, too small for appreciable light scattering at low incident light intensity. These small unilamellar vesicles (SUVs), which are crude models of cell membranes, are energetically unstable and aggregate spontaneously to form multilamellar vesicles (MLVs), as shown in Figure 2-9-5. MLVs range from 50 to 500 nm in diameter and are effective light scatterers.

The conversion of SUVs to MLVs can be monitored using laser light scattering. As the SUVs convert to MLVs, the intensity of light scattering increases. Because of the complexity of this system, no quantitative analysis of the light-scattering data can be made. For instance, whether the increased light scattering is caused by increased number density of the MLVs, increased size of the MLVs, or a combination of both is not clear. The data can, however, be used to obtain a rough estimate of the rate of conversion of SUVs to MLVs. Most of the plots we obtained of scattered light intensity vs. time showed a short induction period, during which the intensity stayed relatively constant, followed by a short period of rapid conversion and a nearly linear increase in light

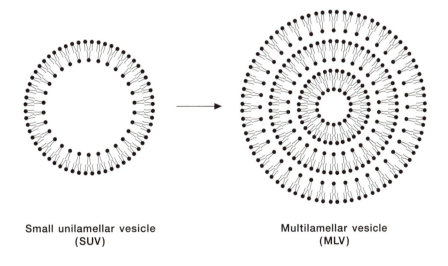

Small unilamellar vesicle   Multilamellar vesicle
        (SUV)   (MLV)

FIGURE 2-9-5. Structures of small unilamellar and multilamellar vesicles.

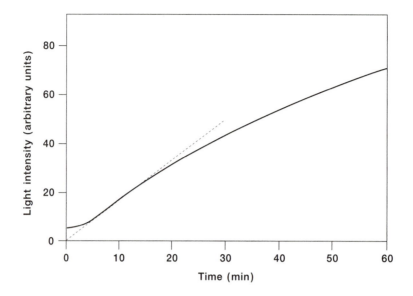

FIGURE 2-9-6. Intensity of light scattered at 90° to a sample of aggregating vesicles at 19.5° C. The "initial rate" of aggregation is taken as the slope of the dashed line.

intensity with time (Figure 2-9-6). We used a ruler to find the "slope" of this latter section and used this slope as a rough measure of the rate of conversion of SUV to MLV. A plot of this "initial rate" temperature (Figure 2-9-7) shows a maximum at about 20° C.

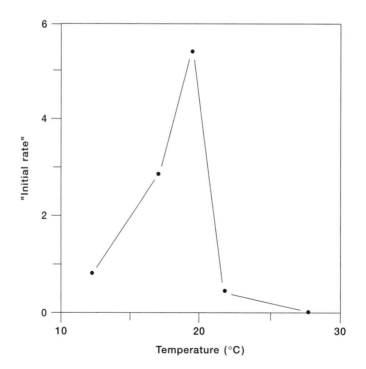

FIGURE 2-9-7. Initial rate of vesicle aggregation as a function of temperature.

The biological relevance of the SUV to MLV conversion is currently unknown, but studies of this type can help determine the conditions that stabilize the unilamellar form of the bilayer. Interesting follow-up experiments might be to vary the pH of the solution, use lipids of different chain lengths and degrees of unsaturation, or attempt to include various types of proteins in the bilayer.

LIGHT SCATTERING AND PARTICLE SIZE

The dependence of light scattering on particle size and concentration is examined, using polystyrene spheres with diameters ranging from 100 to 1,000 nm. Because of the size of the spheres, Rayleigh theory is inadequate to describe the light scattering, and we introduce Rayleigh-Debye theory to provide a more accurate model of the intensity of scattered light.

DEGREE OF DIFFICULTY

Experimental: difficult
Conceptual: difficult

MATERIALS

- red laser
- green laser *(optional)*
- light-detection system
- cuvette holder
- beam splitter
- cuvette, 1 cm^2
- sample holder
- syringe, 30 mL
- 0.22-μm filter
- 20 vials, 4 mL
- pipettes
- latex spheres (107, 343, 778, and 1,052 nm) (Duke Scientific, Palo Alto, CA, or Sigma Chemicals, St. Louis, MO)

PROCEDURE

1. Prefilter distilled water with a 0.22-μm filter. Rinse the cuvette and vials with filtered distilled water.

2. Prepare suspensions of latex spheres of various concentrations in dust-free water. The suspensions we used for 90° scattering had the following concentrations (spheres/mL):

    a. 107-nm spheres: 2.51×10^7, 3.16×10^8, 3.16×10^9
    b. 343-nm spheres: 7.94×10^5, 7.41×10^6, 6.31×10^7
    c. 778-nm spheres (Sigma L8 spheres): 8.91×10^4, 1.00×10^6, 1.26×10^7
    d. 1,052-nm spheres (Sigma L11 spheres): 3.16×10^4, 2.51×10^5, 3.98×10^6

These solutions are prepared by making serial dilutions of the stock microsphere solutions volumetrically.

68

3. Set up the apparatus as shown in Figure 2-4-1. Position the light detector to measure scattering at 90°. Measure the background light scattering intensity using a sample of dust-free water.

4. Take readings of light intensity for each of the solutions of latex spheres at 90°. Keep the temperature of the solutions as constant as possible. Monitor the laser power using the beam splitter and take measurements only when the power is at some predetermined value. Always start with the most dilute solution and end with the most concentrated solution for each particle size. Carefully rinse the cuvette with distilled water when switching from one particle size to the next.

5. *(Optional)* Make light-intensity measurements for various sphere sizes and concentrations at 180°.

6. *(Optional)* Repeat the experiment using a green laser.

HAZARDS AND PRECAUTIONS

Wear eye protection throughout the experiment.

DISPOSAL

The latex spheres solutions may be washed down the sink.

DISCUSSION

The Rayleigh model of light scattering, derived in the introduction to this chapter, predicts that scattering intensity should be proportional to the concentration of scattering centers and the square of the volume of the particles. Further, light scattered at 180° should be twice as intense as light scattered at 90°. In this experiment, however, the latex spheres are not very small compared to the wavelength of laser light used, and Rayleigh theory would not be expected to accurately model the scattering for all of the particle sizes. Rayleigh-Debye theory (closely related to Mie theory) provides a better description of the light scattering observed in this experiment, because it accounts for the interference effects of multiple scattering centers on the same particle and the correspondingly more complex angular dependence of the scattering. In particular, Rayleigh-Debye theory introduces a dimensionless form factor $P(\theta)$ into the Rayleigh relationship to account for the finite size of the particles. The form factor can be thought of as a correction factor to Rayleigh theory, and the scattering intensity predicted by Rayleigh-Debye theory (I_{RD}) is related to the intensity predicted by Rayleigh theory (I_R) according to

$$I_{RD} = I_R P(\theta) \qquad (2\text{-}10\text{-}1)$$

In the limit of low-angle scattering ($\theta \approx 0$) or very small spheres ($V \rightarrow 0$), the function $P(\theta)$ goes to unity so that

$$I_{RD} = I_R \qquad (2\text{-}10\text{-}2)$$

For perfect spheres, when $\theta > 0$, Kerker (1) gives an expression for the form factor $P(\theta)$ that is easily calculated from the scattering angle θ, the sphere radius a (nm), and the refractive index of the medium, n:

$$P(\theta) = \frac{3(\sin u - u \cos u)}{u^3} \tag{2-10-3}$$

where

$$u = 2ka \sin\left(\frac{\theta}{2}\right) \tag{2-10-4}$$

and

$$k = \frac{2n\pi}{\lambda} \tag{2-10-5}$$

where the wavelength λ is expressed in nanometers. The refractive index of water is approximately 1.33 at room temperature. Values of $P(\theta)$ for the 632.8-nm light of the He–Ne red laser at a scattering angle of 90° are listed in Table 2-10-1. Note that the value of $P(\theta)$ decreases with increasing particle diameter, indicating an increasing deviation from Rayleigh theory.

The data from this experiment can be used to investigate the dependence of scattered light intensity on particle size and concentration. Plots of scattered light intensity at 90° vs. particle concentration are shown in Figures 2-10-1 and 2-10-2 for the 107-nm and 778-nm spheres, respectively. Also presented are plots of the logarithm of the scattered light intensity against the logarithm of the particle concentration, which should yield a straight line with a slope of 1 if the relationship between scattering intensity and concentration is linear. As predicted by Rayleigh-Debye theory, the 107-nm spheres showed a linear relationship between particle concentration and scattered light intensity over the entire range of concentrations studied. The data for the 778-nm spheres show significant deviation from the predictions of Rayleigh-Debye theory at high particle concentrations (Figure 2-10-2). Specifically, the three concentrations of spheres lower than 2×10^7 spheres/mL demonstrated linear relationships between scattered light intensity and concentration, but the scattered light intensity for the suspension of concentration 1.58×10^8 spheres/mL was lower than predicted by Rayleigh-Debye theory. The data for light intensity at 180° (Figure 2-10-3) are more difficult to interpret because the light intensity observed is a combination of transmitted light (light that did not scatter off any particle) and light that happened to scatter at 180°.

The volume dependence of the scattered light intensity can be studied by interpolating or extrapolating the values of scattered light intensity to some fixed concentration for each particle size. In other words, graphs of intensity vs. concentration can be used to find the intensity of scattered light that would be expected at some fixed concentration for each particle size. These values can then be used to determine how accurately Rayleigh and Rayleigh-Debye theory model the volume dependency of scattered light intensity for the spheres used in the experiment. Keep in mind that Rayleigh theory predicts that the scattered light intensity should be proportional to V^2 and Rayleigh-Debye theory predicts that it should be proportional to $P(\theta)V^2$.

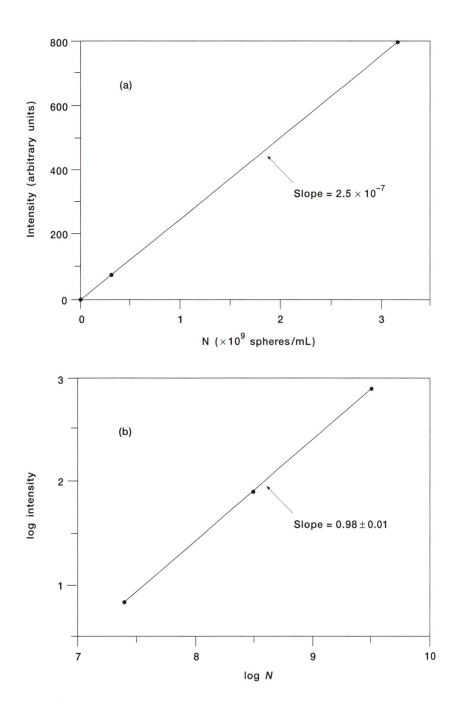

FIGURE 2-10-1. Plot **a** represents the intensity of light scattered at 90° vs. particle concentration for the suspension of 107-nm latex spheres, and **b** is a plot of the logarithm of the scattered light intensity vs. the logarithm of the particle concentration.

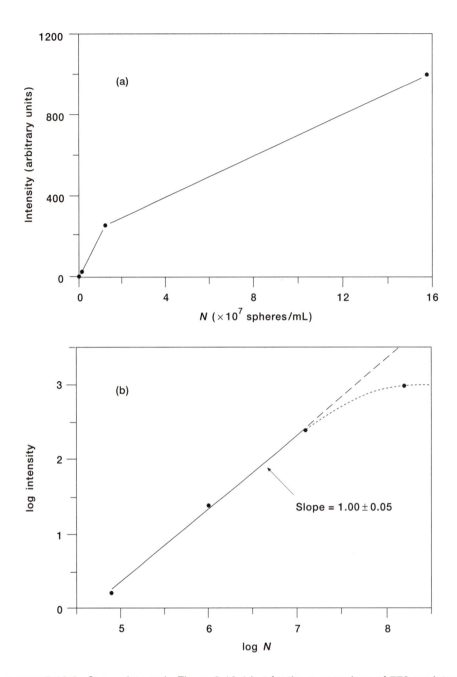

FIGURE 2-10-2. Same plots as in Figure 2-10-1 but for the suspensions of 778-nm latex spheres.

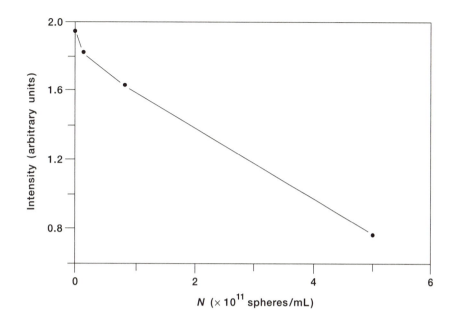

FIGURE 2-10-3. A plot of the intensity of light scattered at 180° vs. particle concentration for the 107-nm latex spheres.

Another way of investigating how accurately the Rayleigh and Rayleigh-Debye theories model the light scattering is to determine (by interpolation or extrapolation) the concentrations of each particle size that correspond to some constant value of scattered light intensity. The values of N, V, and θ can then be used to determine how accurately Rayleigh and Rayleigh-Debye theories predict the intensity of scattered light, as shown in Table 2-10-1. If the Rayleigh theory is accurate, then the value of NV^2 should be constant for constant light scattering intensity. If Rayleigh-Debye theory is accurate, then the value of $P(\theta)NV^2$ should be constant for constant light-scattering intensity. Not surprisingly, Table 2-10-1 shows Rayleigh-Debye theory to be much more accurate for these particle sizes.

An interesting follow-up study might be to examine the spatial distribution and polarization of the scattered light for various sphere sizes.

TABLE 2-10-1. LIGHT SCATTERING AT 90° FOR POLYSTYRENE SPHERES[a]

D	V	N	$P(\theta)$	NV^2	$P(\theta)NV^2$
1,052	924	154,881	3.21×10^{-4}	1.32×10^{11}	4.24×10^{7}
778	374	537,032	1.70×10^{-3}	7.51×10^{10}	1.28×10^{8}
343	32	3,715,352	1.24×10^{-2}	3.80×10^{9}	4.70×10^{7}
107	1	33,884,416	0.741	3.39×10^{7}	2.51×10^{7}

[a] D indicates particle diameter (nm); V, volume in relative units referenced to the 107-nm particle volume as unity; N, number of spheres; and $P(\theta)$, the dimensionless form factor of Raleigh-Debye theory.

REFERENCE

1. Kerker, M. *The Scattering of Light and Other Electromagnetic Radiation.* Academic; New York: 1969; 417.

CHAPTER 3

DIFFRACTION: LIGHT SCATTERING FROM ORDERED SYSTEMS

This chapter presents four experiments that use lasers to demonstrate diffraction effects. Three of these experiments provide insight into X-ray crystallography. Experiment 3-2 uses a laser to generate diffraction patterns from optical transforms, slides with dot patterns in regular arrays analogous to crystalline structures. Experiments 3-3 and 3-4 examine diffraction of laser light from two- and three-dimensional crystals composed of latex spheres. The remaining experiment, experiment 3-1, provides a simple demonstration of diffraction effects using a periodically ruled surface, such as a Vernier caliper. This experiment is particularly useful as an introduction to diffraction effects.

The chapter's introduction provides a concise overview of the theory of diffraction, with an emphasis on X-ray crystallography. All the experiments require an understanding of the topics dicussed next under "Basic Theory of Diffraction." The material under "Diffraction from Crystals" is needed to understand experiments 3-2, 3-3, and 3-4. At the end of the introduction, we present a brief introduction to holography and a guide to the literature for making holograms with a low-power helium–neon laser. We placed this treatment of holography in this chapter because a hologram may be regarded as a large collection of microscopic diffraction gratings.

BASIC THEORY OF DIFFRACTION

The phenomena of diffraction and light scattering are closely related. As discussed in Chapter 2, light scattering is one means by which light can interact with matter. In the process of light scattering, the electric field of light interacts with the electrons in a particle, causing them to oscillate with the same frequency as the incident light. The particle in this state is known as an oscillating dipole and reradiates light of the same frequency as the incident light in almost all directions. This reradiated light is called scattered light.

As we alluded in Chapter 2, when more than one scattering particle is present, interference must be taken into consideration. Interference arises from the wavelike properties of light. All electromagnetic radiation has an associated wavelength λ (m) and frequency ν (Hz $= \sec^{-1}$), such that their product (in a vacuum) equals the speed of light c (m/s):

$$c = \lambda \, \nu \qquad\qquad (3\text{-}1)$$

When electromagnetic radiation from several sources overlap in space, the individual waves add together. Consider two waves that have the same amplitude and wavelength. If the individual waves are in phase, meaning that their maxima and minima line up in space at the same time, they reinforce each other, and *constructive interference* results (Figure 3-1). The amplitude of the resulting wave is twice that of the interfering waves. Conversely, when the waves are separated by half a wavelength (completely out of phase), so that the maximum amplitude of one occurs simultaneously with the minimum amplitude of the other, the two waves cancel each other, resulting in a wave of zero amplitude. This interference is *destructive interference*.

A simple example of how interference effects change the spatial distribution of scattered light is that of two oscillating dipoles. Imagine a scattering experiment in which two particles, separated by a distance d (m), are irradiated by a beam of light of just one frequency, such as one from a laser. As in the derivation of Rayleigh scattering in Chapter 2, we will assume that the particles are small enough to be considered point sources and that they reradiate light of the

Constructive interference

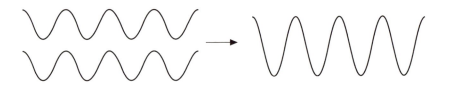

Destructive interference

FIGURE 3-1. Constructive and destructive interference between two waves with the same wavelength and amplitude.

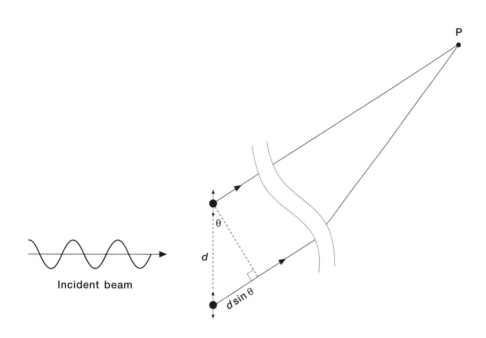

FIGURE 3-2. Two oscillating dipoles (point sources) aligned perpendicularly to an incident beam. The expression $n \lambda = d \sin \theta$ arises from the difference in distance from the two dipoles and the distant point P.

same frequency as the incident light. If they are positioned as shown in Figure 3-2, they will also be in phase, that is, they will oscillate together in time.

At any point in space, the light scattered from the two particles will interfere either constructively or destructively, depending on whether the light coming from the two sources is in phase or out of phase at that particular point. The only time that the light will be completely in phase is if the distance from the two dipoles is identical, or is different by some integral multiple of the wavelength of light. Figure 3-2 shows that for any distant point P, the difference in pathlength is equal to $d \sin \theta$, where θ is the angle formed between the point P and each of the scattering centers, with respect to the axis x (the angle is the same for both scattering centers because of the assumption that P is very far away). Hence, the condition for constructive interference is

$$d \sin \theta = n \lambda \qquad (3\text{-}2)$$

where n is any positive integer.

This case of two point sources is the simplest example of diffraction. Similar arguments can be made to account for the interference patterns that result from scattering from any number of particles, although the problems become increasingly more complex, because the amplitudes of the waves from all of the different sources must be added together. When large numbers of particles are present, interference patterns (diffraction effects) are usually observed only when the particles have a regular spatial ordering. If the particles are randomly dispersed, as in the experiments in Chapter 2, then the interference effects among the light scattered by the various particles tend to cancel each other. At any point outside the sample, the intensity of light observed is approximately equal to the number of particles times the intensity of light that would be observed from just one particle. When the particles do have a regular order, however, the waves of light interfere destructively at almost all locations, except for a few spots or rings of constructive interference.

DIFFRACTION FROM CRYSTALS

Crystals are an example of an array of scattering centers that have a regular order, and for this reason crystalline solids diffract light. The diffraction of light from crystals is an extremely important phenomenon in chemistry because the resulting diffraction patterns provide detailed information about the regular three-dimensional arrays of atoms or molecules in the crystals. In this chapter, we examine two important diffraction phenomena observed in studies of crystals. The first is known as *Kossel rings,* which are observed only if the incident light strongly induces scattering dipoles in the crystal. These rings are often weak in X-ray diffraction studies of crystals, and for this reason we omit discussion of this diffraction effect for now. Experiment 3-4 provides an example of Kossel diffraction patterns using a colloidal crystal irradiated with visible light.

The dominant diffraction effect observed in studies of crystals is a pattern of dots generally referred to as a *Bragg diffraction pattern.* A simple way of understanding the Bragg diffraction pattern is to imagine that the planes of scattering centers in the crystal reflect some of the incident electromagnetic radiation passing through the crystal, just like a plane of glass transmits most of the light but reflects a small portion. Bragg diffraction results from the interference among rays of light reflected from parallel planes in the crystal. Figure 3-3 illustrates the conditions for constructive interference from one set of scattering planes. When the two rays of light reach the distant point P, the lower ray will have traveled $2\,d\,\sin\,\theta$ farther. The rays of light will interfere constructively only if this difference in distance is equal to an integral multiple of the wavelength of light. The corresponding condition for constructive interference is known as Bragg's law,

$$2\,d\,\sin\,\theta = n\,\lambda \qquad\qquad (3\text{-}3)$$

named for William Henry Bragg (1862–1942) and Sir William Laurence Bragg (1890–1972), father and son British physicists who shared the 1915 Nobel Prize in physics for their work in X-ray diffraction and crystallography.

Bragg's law predicts that only certain wavelengths of light will produce a diffraction pattern from a crystal. Bragg's law can be rearranged to

$$\sin\,\theta = \frac{n\lambda}{2d} \qquad\qquad (3\text{-}4)$$

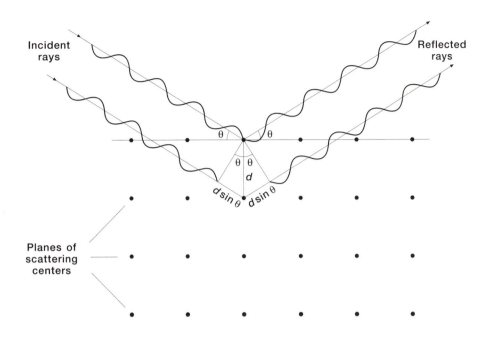

FIGURE 3-3. Physical basis of Bragg diffraction.

The value of $\sin \theta$ can be no greater than 1, which implies that the value of $n\lambda/d$ can be no greater than 2:

$$\frac{n\lambda}{d} \leq 2 \qquad (3\text{-}5)$$

Because d, the spacing between scattering centers, is very small for a crystalline lattice, the wavelength of light used to perform the scattering experiment must be correspondingly very small to observe Bragg diffraction. Specifically, the wavelength of light must be approximately the same length or even smaller than d. For most crystals, d is approximately 0.1 nm, and electromagnetic radiation with wavelengths of 0.1 nm or smaller corresponds to X-ray radiation. The colloidal crystals prepared from latex spheres in experiments 3-3 and 3-4 have much larger lattice spacings, and diffraction patterns can be observed with visible light.

Figure 3-4 depicts a typical experimental setup for X-ray diffraction studies, and Figure 3-5 shows an example of a Bragg diffraction pattern (1). In many X-ray diffraction experiments, the diffraction patterns are recorded on photographic film, and the (x,y) coordinates of the spots are used to generate a physical model of the crystalline material. X-ray diffraction studies elucidate the structure of the unit cell of the crystal, the fundamental three-dimensional assembly of atoms or molecules that is repeated in all dimensions to form the complete space lattice of the crystal. Figure 3-6 shows the unit cell structures for each of the seven basic crystal types.

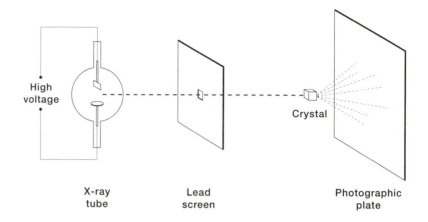

High
voltage

X-ray
tube

Lead
screen

Crystal

Photographic
plate

FIGURE 3-4. Schematic representation of the setup for an X-ray diffraction experiment.

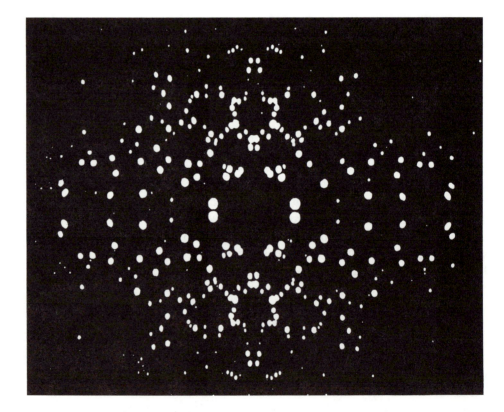

FIGURE 3-5. Bragg diffraction pattern (1) from a crystal of $Zn(cyclam)(ClO_4)_2$.
Photo courtesy of K.O. Hodgson.

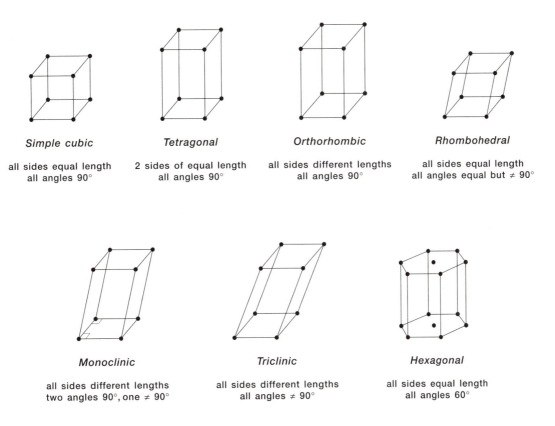

FIGURE 3-6. Unit cell structures of the seven basic crystal types.

An important consideration in interpreting diffraction patterns is that crystals are never perfect periodic point lattices. In addition to defects in the crystal structure, the atoms or molecules that comprise the crystal are not stationary but oscillate about their minimum potential energy positions. These motions tend to blur the diffraction spots that otherwise would be quite sharp. This type of blurring is observed in the diffraction patterns from the two-dimensional crystals composed of latex spheres studied in experiment 3-3. In addition, "extra" spots are sometimes observed in diffraction patterns. These spots arise from the interaction of the incident light with *phonons*, which are elastic waves of optical and acoustical frequencies that travel through the lattice (2).

In practice, X-ray diffraction experiments are too dangerous and costly to be performed in student laboratories. Conceptually similar experiments can be prepared, however, that demonstrate diffraction effects using visible light. The experiments we present use latex crystals and optical transforms as media that diffract visible light. The monochromaticity and coherency of laser light makes this light ideal for such diffraction experiments, producing clean and intense diffraction patterns.

HOLOGRAPHY

A hologram can be considered a large collection of microscopic diffraction gratings. Central to diffraction phenomena are interference effects in which light from some source or reflected from some object is divided into two or more waves that traverse different paths before recombining at the same spot. Upon recombination, the intensity of the light at the spot is proportional to the square of the sum of the electric-field amplitudes of the different light waves. The intensity is a maximum when the phases of the different recombining waves reinforce each other (are in step) and a minimum when the phases of the different recombining waves cancel each other (are out of step). A particularly striking use of interference effects is *holography,* the method for recording information from a three-dimensional object in such a way that a three-dimensional image can be reconstructed. Holography was developed before the laser; the first hologram was recorded by Gabor (3) in 1948, who received the 1971 Nobel Prize in physics for this discovery. Because holography exploits interference effects, it requires light with a high degree of spatial and temporal coherence (phase registry). Consequently, the use of holography has taken off with the advent of the laser. Today, holograms are commonly found on magazine covers, breakfast cereal boxtops, credit cards, and the paper currency of some countries (the latter two examples are simple means to make counterfeiting more difficult).

When you view a hologram, the image appears as the real object. You can look around corners, see inside crannies and crevices, and so forth. A single hologram carries more information about the three-dimensional nature of the pictured object than does a series of photographs taken at different angles. Yet the resolution of features in a hologram is limited. Only features that are comparable to or larger than the wavelength of the light used to make the hologram can be pictured. Once coherent light sources are able to be extended into the X-ray portion of the electromagnetic spectrum, X-ray holography should revolutionize the way structural information can be obtained on molecules and molecular assemblies. Until then, we are limited to creating holograms using coherent visible light to show the three-dimensional structure of macroscopic objects.

Holography of a real object is a two-step optical imaging process. First, an interference pattern is recorded on a photographic emulsion to produce the hologram. Second, the hologram is illuminated with light (which need not be the same as the recording light) such that the transmitted or reflected light yields a reconstructed image of the original object recorded. Figures 3-7a and 3-7b illustrate the basic principles necessary to make a hologram. Photographic film is exposed simultaneously to waves of light scattered from an opaque object and to waves of light from a reference source (Figure 3-7a). The two sets of waves produce an interference pattern on the photographic film, which is recorded by the photographic emulsion to form a hologram. The film is processed and then illuminated with only the reference beam present (Figure 3-7b). Most of the light from the reference beam passes through the hologram. Some of it, however, is diffracted by the interference pattern in the photographic film. The construc-

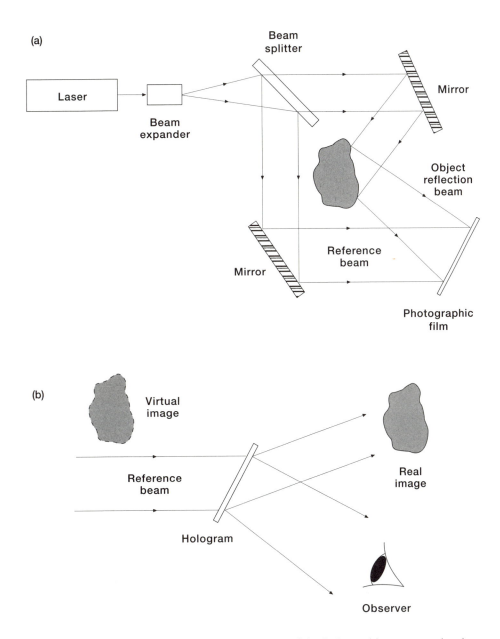

FIGURE 3-7. Panel **a** is a schematic representation of the holographic process, showing the recording of a hologram produced by the interference of the wave fronts from an object and a coherent reference beam. Panel **b** presents the hologram readout, in which the image of the object is reconstructed.

tive interference of the diffracted waves reconstructs the original pattern of light waves scattered from the object so that to an observer, the waves appear to be coming from the object itself. These waves constitute the *virtual image*. Just as a diffraction grating gives diffracted orders on either side of the zero-order straight-through position, the hologram generates a second image. This image, which is usually inferior in quality to the virtual image, is called the *real image*.

The hologram serves to provide a "window" on the object illuminated by the laser—a window through which the object may be viewed at different angles.

Figure 3-7 shows only one way to prepare a hologram. Many other experimental arrangements can be used. One is a 360° hologram, in which the window for viewing the object wraps all the way around it (4). Another is an unsplit-beam setup, in which the same beam is used twice by placing the object to be recorded directly behind the film so that the laser beam passes through the semitransparent film, thereby exposing the film with the reference beam, strikes the object, and reflects back to the film as the scattered object beam (5). Soon after these principles of holography were understood came the realization that computer programs could be designed to generate holograms.

Note that a hologram appears as a crazy-quilt collection of interference fringes that do not resemble the real object but contain information on the phase and amplitude distribution of the scattered light waves from the recorded object. These interference fringes act as numerous microscopic diffraction gratings; variations in the contrast and the spacings between interference fringes produce variations in the intensity of the read-out beam from the hologram. The film used in a photograph and hologram work in the same way; both involve the conversion of silver halide grains in the film emulsion to silver atoms. The density of silver atoms is greatest where the light intensity is greatest. Gabor's discovery was that by superimposing the (coherent) light scattered from the object with the unscattered (coherent) light reference beam, the intensity pattern will store information about the phase and amplitude of the scattered light from the object that subsequently can be read out.

Because holography is based on interference effects, all distances must be held fixed to better than one quarter of the wavelength of the light used to record the hologram. Most of the failures in hologram production can be traced to failures in mechanical stability and vibration isolation. Fortunately, inexpensive relatively large, vibration-free platforms (optical tables) can be constructed and good-quality holograms can be made using a low-power helium–neon laser (4,5) (but not with most red diode lasers because of less coherent output) combined with fast, high-resolution film. Indeed, reports exist of high school classes routinely succeeding in making holograms (5). Moreover, several companies offer kits for the amateur holographer (e.g., 6–8). To minimize the requirements for mechanical stability, shorter exposure times are preferred. Consequently, the helium–neon laser of choice should have an output of several milliwatts, although acceptable holograms are still obtainable from helium–neon lasers of approximately 1-mW power. Rather than rating any one holographic setup or procedure as best, we urge interested readers to consult references 4–11. The production of holograms not only illustrates the phase information carried by coherent light scattered from an object, but it also provides an excellent introduction to the principles behind photography.

REFERENCES

1. Tyson, T.A., Hodgson, K.O., Hedman, B., Clarke, G.R. *Acta Crystallogr., Sect. C: Cryst. Struct. Commun.* C46 (1990) 1638–1640.

2. Besançon, R.M. *The Encyclopedia of Physics,* 3rd ed. van Nostrand Reinhold; New York: 1985; 285–288.

3. Gabor, D. *Nature* 161 (1948) 777.

4. Kallard, T. *Exploring Laser Light.* Optosonic. Reprinted by American Association of Physics Teachers, 5110 Roanoke Pl., Ste. 101, College Park, MD 20740. 1977 (See p. 221–259 for a clear, extensive discussion of holography.)

5. Altman T.C. *Phys. Teach.* 30 (1992) 220, and references therein.

6. *The 1994 Newport Catalog.* Newport Corp., 1791 Deere Ave., Irvine, CA 92714: 1994; 15.1–15.12.

7. Scientific Laser Connection of Arizona, 5021 N. 55th Ave., Ste. 10, P.O. Box 433, Glendale, AZ 85311.

8. Metrologic Instruments, Inc., Blackwood, NJ 08012.

9. Stroke, G.W. *An Introduction to Coherent Optics and Holography.* Academic; New York: 1969.

10. Smith, H.M. *Principles of Holography,* Wiley; New York: 1975.

11. Wilson, J., Hawkes, J.F.B. *Optoelectronic: An Introduction.* Prentice Hall; Englewood Cliffs, NJ: 1983.

DIFFRACTION FROM A VERNIER CALIPER

A Vernier caliper is used to generate a diffraction pattern with a laser (1). The spacing of the diffraction spots permits calculation of the wavelength of the laser light.

DEGREE OF DIFFICULTY

Experimental: easy
Conceptual: moderate

MATERIALS

- laser
- Vernier caliper with engraved main scale
- meter stick
- mirror (optional)

PROCEDURE

1. Place the Vernier caliper on a table three to four meters from a wall. Direct the laser beam to be perpendicular to the engraved lines and to graze the main scale of the caliper at an angle of approximately 3° (Figure 3-1-1). A mirror may be helpful for directing the laser beam.

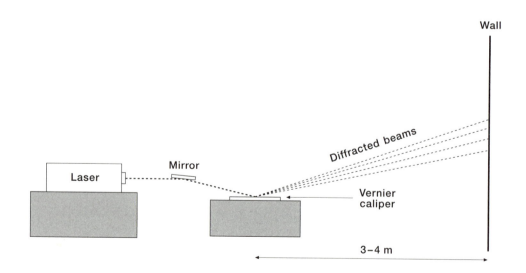

FIGURE 3-1-1. Schematic diagram of experimental setup for observing a diffraction pattern from the engraved scale on a Vernier caliper.

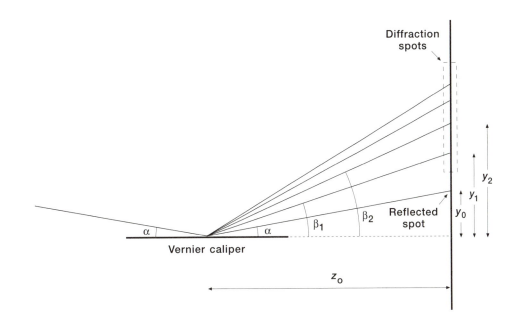

FIGURE 3-1-2. Definitions of the variables α, β_1, β_2, ... , y_0, y_1, y_2, ... , z_0.

2. Measure the vertical positions of the reflected spot and the diffraction spots on the wall relative to the vertical position of the caliper (Figure 3-1-2), that is, measure y_0, y_1, y_2, etc. A convenient method of determining the vertical position of the caliper on the wall is to mark the point where the laser beam would strike the wall in the absence of the caliper. The vertical position of the caliper on the wall is the point midway between this spot and the spot corresponding to the reflected laser beam. Also measure z_0, the horizontal distance between the wall and the point where the laser beam strikes the caliper.

HAZARDS AND PRECAUTIONS

Never look directly into the laser beam. Set up the experiment to minimize the risk of directly viewing the reflected and diffracted beams.

DISPOSAL

None.

DISCUSSION

When a laser beam shines on a Vernier caliper scale, the surface of the caliper reflects most of the laser light, but light that strikes the engraved markings is scattered in all directions (Figure 3-1-3). The light scattered from two of the engravings interferes constructively only in certain locations in space, resulting in a diffraction pattern. Note that the origin of this diffraction pattern is conceptually similar to that of two point sources discussed in the introduction to this chapter. The locations of the regions of constructive

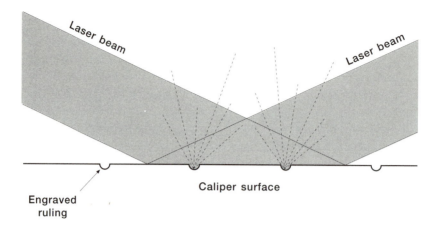

Engraved
ruling

Caliper surface

FIGURE 3-1-3. Laser light scattered by engravings on the caliper surface.

interference can be determined by considering the difference in pathlength between two beams of light that strike adjacent engraved markings on the caliper.[1] If the difference in pathlength is an integral multiple of the wavelength, then the two waves interfere constructively. In Figure 3-1-4, if point P is assumed to be very far away, then the difference in pathlength between the two beams of light will be $d \cos \alpha - d \cos \beta$. Thus, the condition for constructive interference is

$$d\,(\cos \alpha - \cos \beta_m) = m\,\lambda \qquad (3\text{-}1\text{-}1)$$

where the subscript m has been added to index the diffraction spots. When $m = 0$, $\alpha = \beta_0$, and the diffraction spot cannot be distinguished from reflection.

According to equation 3-1-1, if the distance d between the engraved markings is known, then the wavelength of light can be calculated from the angles α and β. In turn, the angles α and β can be determined from the vertical positions of the diffraction spots using trigonometry. Figure 3-1-5 shows the relevant geometry for one of the diffraction spots. The hypotenuse of the triangle is given by

$$\sqrt{z_o^2 + y_m^2} \qquad (3\text{-}1\text{-}2)$$

and

$$\cos \beta_m = \frac{z_o}{\sqrt{z_o^2 + y_m^2}} \qquad (3\text{-}1\text{-}3)$$

Similarly, for the reflected spot,

$$\cos \alpha = \frac{z_o}{\sqrt{z_o^2 + y_o^2}} \qquad (3\text{-}1\text{-}4)$$

1. Sirohi (1) provides a somewhat different mathematical treatment of this experiment, which begins with the grating equation and employs a Taylor expansion.

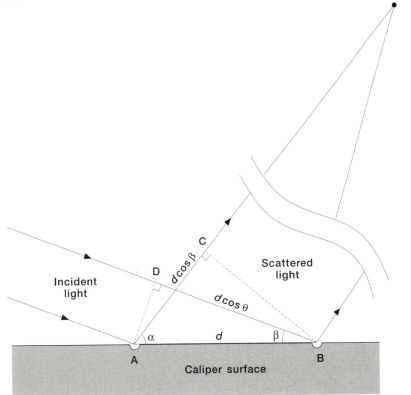

FIGURE 3-1-4. Physical basis of the diffraction equation $d(\cos\alpha - \cos\beta_m) = m\lambda$.

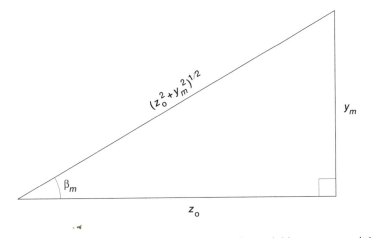

FIGURE 3-1-5. Geometric relationships among the variables y_m, z_o, and β_m.

In theory, each diffraction spot could be used to determine the wavelength of light, but in practice it is more accurate to use several spots and average the results. Sirohi (1) reports an uncertainty of less than 1% for the determination of the wavelength of light of a He–Ne laser in an experiment identical to the one described here.

REFERENCE

1. Sirohi, R.S. *A Course of Experiments with He–Ne Laser.* John Wiley; New York: 1985; 45–48.

OPTICAL TRANSFORMS

Optical transforms are two-dimensional arrays of dots or other patterns that produce diffraction patterns with visible light. The diffraction patterns produced reflect the symmetry, size, and spacings of the optical transform patterns, illustrating the usefulness of X-ray diffraction in determining crystal structure.

DEGREE OF DIFFICULTY

Experimental: easy
Conceptual: easy

MATERIALS

- red laser
- green laser *(optional)*
- Optical Transform Kit or suitable 35-mm slides of dot arrays (see step 1)
- ruler

PROCEDURE

1. Those who wish to prepare their own optical transform materials can do so by drawing dot arrays with a computer program and then using standard copy and photographic techniques to prepare 35-mm slides (1) so that dot centers are separated by microns. Alternatively, the Institute for Chemical Education (ICE) at the University of Wisconsin offers a complete optical transform kit at modest cost that includes the necessary materials and instructions for simulated diffraction studies (2,3). The dot arrays available on 35-mm slides from ICE are shown in Figure 3-2-1.

2. Position the laser a few meters away from a screen or blank wall, as shown in Figure 3-2-2. Darken the room. Hold the optical transform slide in the laser beam approximately halfway between the laser and the screen.

3. Observe the diffraction patterns created by each of the dot arrays. Having an enlarged picture of the dot arrays on hand may be helpful for comparison with the diffraction patterns. The diffraction patterns produced by the dot arrays in Figure 3-2-1 are shown in Figure 3-2-3.

4. Project the laser beam directly onto the screen, and view the spot on the screen through the optical transform slide. You should observe the same diffraction patterns as before (in a sense, the diffraction patterns form directly on your retina).

5. *(Optional)* Observe the diffraction patterns produced by laser light of a different wavelength, such as a green He–Ne laser. Green and red laser light produce clearly different pattern sizes. Verify that this difference should be true according to the theory of Fraunhofer diffraction as discussed below.

92

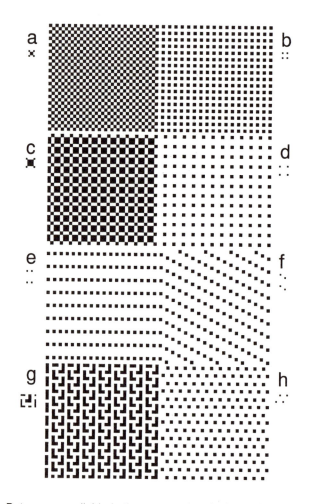

FIGURE 3-2-1. Dot arrays available in the prepared optical transform kit (2). Depicted alongside the arrays are the structures of each two-dimensional unit cell. From (3), with permission.

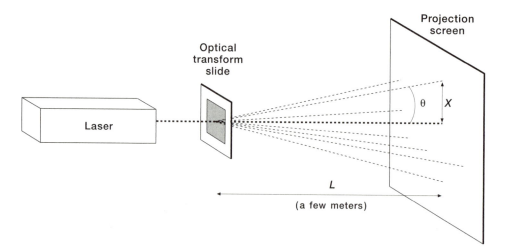

FIGURE 3-2-2. Schematic diagram of experimental setup for observing diffraction patterns from optical transforms. X is the distance from the point where the laser beam would hit the screen in the absence of the optical transform to the point of a diffraction spot. L is the distance between the slide and the screen.

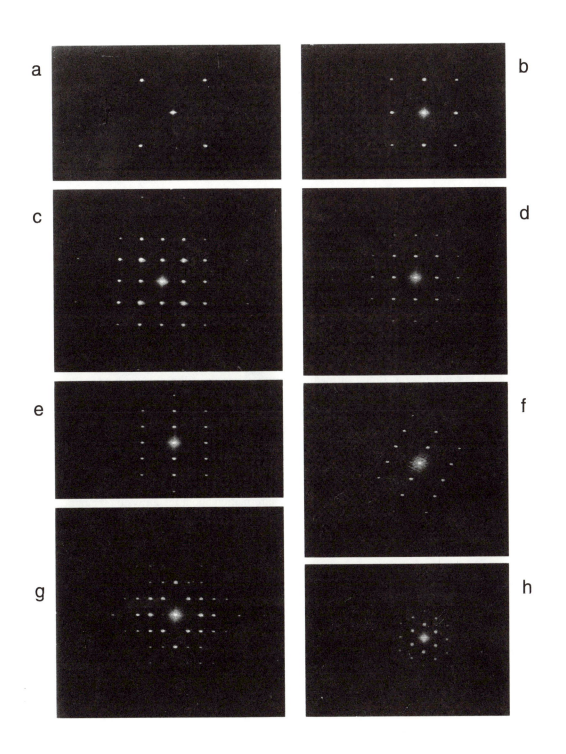

FIGURE 3-2-3. Diffraction patterns produced by the dot arrays shown in Figure 3-2-1. From (3), with permission.

HAZARDS AND PRECAUTIONS

Never look directly into the laser beam.

DISPOSAL

None.

DISCUSSION

As discussed in the introduction, crystal lattices diffract X-ray radiation because the spacing between the lattice points is on the same order as the wavelengths of X-ray radiation. The resulting diffraction patterns yield information on the structure and spacing of the crystal lattice. Optical transforms can be used to simulate such studies. The spacings between the dots on the optical transform slides are several orders of magnitude greater than the dimensions of a unit cell of a typical crystal. In fact, the spacings are on the order of the wavelengths of visible light, and optical transforms diffract visible light in much the same way crystal lattices diffract X-ray radiation.

One key difference between optical transform diffraction and X-ray crystallography is that crystal lattices are three dimensional, whereas optical transforms are two dimensional. For this reason, Bragg diffraction, which arises from interference between rays of light reflected from parallel planes in a three-dimensional lattice, does not occur with optical transforms. The model of diffraction appropriate for optical transforms is known as *Fraunhofer diffraction,* which is depicted along with Bragg diffraction in Figure 3-2-4.

Because of the complexity of the arrays on the optical transforms, rigorously modeling the diffraction from these arrays would be quite difficult. The diffraction patterns do, however, provide considerable qualitative information about the arrays on the optical transforms. The optical transform arrays can be considered, just as crystal lattices are, to be composed of unit cells, the smallest unit of the pattern that repeats in all directions to form the array. Even though the shape of the unit cell may be complex, in general the spacings between the diffraction spots reflect the unit cell size according to the same equation that describes the interference between two point sources:

$$d \sin \theta = n \lambda \qquad (3\text{-}2\text{-}1)$$

(see the introduction to this chapter for the derivation of this expression). A useful way of analyzing the diffraction patterns is to measure the distances of various spots from the central (most intense) dot, where the laser beam would hit the screen in the absence of the optical transform. If we call this distance X, and the distance between the slide and the screen L (as shown in Figure 3-2-2), then

$$\tan \theta = \frac{X}{L} \qquad (3\text{-}2\text{-}2)$$

Because $\tan \theta \approx \sin \theta$ for small angles, $\sin \theta = X/L$ can be substituted into the diffraction condition $d \sin \theta = n \lambda$ to obtain

$$\frac{dX}{L} = n \lambda \qquad (3\text{-}2\text{-}3)$$

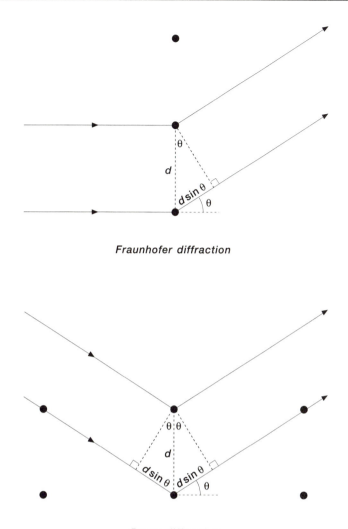

Fraunhofer diffraction

Bragg diffraction

FIGURE 3-2-4. Comparison of Bragg and Fraunhofer diffraction.

Note that the distance between the diffraction spots is directly proportional to the wavelength of light and inversely proportional to the spacing between the dots on the optical transform. This relationship explains why an optical transform with smaller dot spacing has a larger diffraction pattern and why a green laser light source gives a smaller diffraction pattern than a red laser.

The patterns in the optical transform kit (2) are designed to be analogous to various types of crystal structure. As in X-ray diffraction experiments, the shape of the unit cell affects the intensities of the diffraction spots and the symmetries observed in the diffraction pattern. The instructions in the kit give qualitative explanations of how the spot patterns relate to the various unit cell shapes.

Diffraction patterns can be produced with a laser from a variety of common objects, such as a bird feather, piece of hair, a fine mesh screen, or even blood cells (1,4–7).

The spacings of the structures that produce the diffraction effects can be estimated in a manner similar to that used in this experiment.

An interesting follow-up experiment would be to observe diffraction patterns from optical transform slides with arrays that are not perfect two-dimensional patterns. For instance, optical masks could be prepared with discontinuities similar to ones commonly observed in crystals, such as vacancies, interstitials, and screw dislocations. Alternatively, Besançon (8) provides some interesting examples of arrays that mimic thermal oscillations and phonons in a crystal lattice, along with the corresponding diffraction patterns.

REFERENCES

1. Slabaugh, W.H., Smith, D. *J. Chem. Ed.* 51 (1974) 207.
2. *Optical Transform Kit.* Available from Optical Transforms, Institute for Chemical Education, Department of Chemistry, University of Wisconsin-Madison, 1101 University Ave., Madison, WI 53706.
3. Lisensky, G.C., Kelly, T.F., Neu, D.R., Ellis, A.B. *J. Chem. Ed.* 68 (1991) 91.
4. Curry, S.M., Schawlow, A.L. *Am. J. Phys.* 42 (1974) 413.
5. Fischbach, F.A., Bond, J.S. *Am. J. Phys.* 52 (1984) 519.
6. Hwa, Y.P. *Phys. Teach.* 17(4) (1979) 258.
7. Kliev, K., Taylor, A.J. *J. Chem. Ed.* 68(2) (1991) 155.
8. Besançon, R.M. *The Encyclopedia of Physics,* 3rd ed. van Nostrand Reinhold; New York: 1985; 285–288.

DIFFRACTION FROM A TWO-DIMENSIONAL CRYSTAL

A red laser beam is passed through a sample of latex spheres layered between two microscope slides, thereby producing diffraction patterns that reflect the two-dimensional structure of the latex spheres.

DEGREE OF DIFFICULTY

Experimental: moderate
Conceptual: moderate

MATERIALS

- red laser
- (x,y)-translator (such as a scavenged microscope stage)
- adjustable mirror
- microscope slides
- 1 mL of glycerol or ethylene glycol
- 5 drops of latex sphere preparation (91-nm diameter), deionized (see experiment 3-4) or untreated (Sigma product #LB-1. Approx. $25 for 1 mL)
- 1 mg of glass beads, 75–150 μm *(optional)* (Sigma product #G4649. Approx. $13 for 10 g. Also available from Duke Scientific, Palo Alto, CA)

PROCEDURE

1. The chance of observing high-quality diffraction patterns can be maximized by deionizing the latex spheres according to the procedure for experiment 3-4. Fair results can also be obtained using untreated latex spheres, however.

2. Apply a film of glycerol or ethylene glycol to a clean microscope slide using the same technique as that for making a blood smear on a slide.

3. Place one drop of the latex spheres solution (91-nm diameter) on the glycerol film. Position a clean microscope slide on top of this slide.

4. Place the assembly on the translator, and use the mirror to direct the laser beam upward through the microscope slides (see Figure 3-3-1), taking care to avoid looking into the laser beam. Start observing the diffraction patterns as soon as possible, because the latex spheres lose their structure over time.

5. Adjust the position of the slide to create cubic and hexagonal Bragg diffraction patterns on the ceiling. High-quality diffraction patterns are often elusive. If no diffraction patterns are found after 2 minutes of searching, prepare another slide and try again.

6. *(Optional)* Repeat the experiment using glass beads instead of latex spheres.

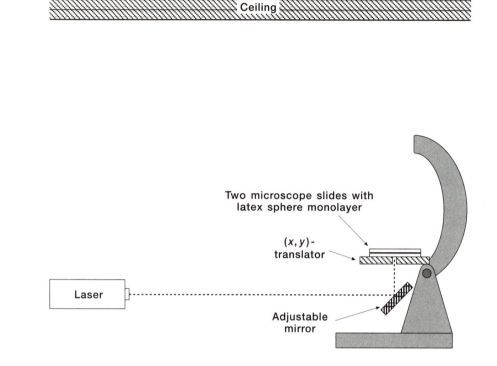

FIGURE 3-3-1. Schematic diagram of experimental setup for observing diffraction patterns from two-dimensional crystals of latex spheres.

HAZARDS AND PRECAUTIONS

Be cautious while observing spot patterns on the ceiling for long periods, as sore necks and eye strain may result. Short observation times present no great problem.

DISPOSAL

No special precautions are necessary.

DISCUSSION

The diffraction patterns observed in this experiment take the form of blotches of light in cubic and hexagonal arrays. The diffraction pattern reflects the ordered structure of the spheres on the microscope slides. The latex spheres are negatively charged and form two-dimensional ordered structures similar to the lattice structure of a crystal; these structures minimize the electrostatic repulsions between spheres. (Please see experiments 2-4 and 3-4 for more information on the interactions between latex spheres.) The inter-sphere distances can be calculated from the spacing of the diffracted spots using the procedure outlined in experiment 3-2.

The two-dimensional structure of the latex spheres is not static but fluctuates fairly rapidly. Some diffraction spots may persist for less than a second and others may display a twitching phenomenon, which suggests that the spheres are somewhat free to move about between the slide surfaces. Over longer periods, the latex spheres in the solution clump together and the diffraction pattern is lost.

The glass spheres produce a diffraction pattern that consists of concentric circles. Note that the microscope slide assembly can also create a series of rings as a diffraction pattern (1), and distinguishing between the two diffraction patterns may not be easy. Unlike the latex spheres, the glass beads are not charged and have essentially no inter-sphere potential. Thus, they would not be expected to form ordered structures like the latex spheres. Rather, the glass beads probably behave much like molecules in a liquid or a gas, moving randomly and bumping into each other but not forming an ordered structure. Not suprisingly, the diffraction pattern of the glass spheres is also similar to the diffraction patterns observed for liquids and disordered solids (2).

REFERENCES

1. Pontiggia, C., Zefiro, L. *Am. J. Phys.* 42 (1974) 692.
2. Hukins, D.W.L. *X-ray Diffraction by Disordered and Ordered Systems.* Pergamon; New York: 1981; 66.

KOSSEL RING DIFFRACTION PATTERNS FROM A COLLOIDAL CRYSTAL

Colloidal crystals composed of latex spheres are prepared that have structures very similar to those of crystals composed of atoms and molecules, except that the collodial crystals have interparticle distances appropriate for diffraction of visible light. Irradiating the crystals with a red laser creates a series of Kossel rings as a diffraction pattern. As an optional exercise, the bulk modulus of the crystal (a measure of its structural strength) can be estimated by making several measurements of the central dark Kossel ring from a large single crystal.

DEGREE OF DIFFICULTY

Experimental: very difficult
Conceptual: difficult

MATERIALS

- red laser
- 0.25-inch-thick plexiglass
- 0.03-inch-thick Teflon sheet
- 30 each: nuts, screws, and washers
- dialysis tubing, 6 inches
- disposable pipettes
- 5 custom sample holders
- (x,y,z)-translator and connector rod
- 5 plastic cuvettes, 1.0 cm^2
- parafilm
- syringe
- converging lens, plano-convex
- latex spheres, 91 nm, 10% solids (Duke Scientific, Palo Alto, CA, or Sigma Chemical Co., St. Louis, MO), 2.0 mL
- 10 g of ion exchange resin, AG 501-X8(D) (Biorad, Richmond, CA)
- conductivity-measuring equipment
- water bath
- ruler
- tracing paper
- Buchner funnels

PROCEDURE

The process of growing the colloidal crystals is fairly lengthy, requiring approximately 2 weeks. Colloidal crystals grow only in solutions with very low electrolyte concentrations. To ensure sufficient deionization of the latex sphere solution, all of these procedures listed below must be followed carefully. Those who cannot generate Kossel ring patterns experimentally may obtain a set of photographs from the American Association of Physics Teachers (1). These photos, however, do not show complete Kossel rings and cannot be used to do the quantitative analysis discussed below.

Regeneration of Ion-Exchange Resin

1. Newly purchased ion-exchange resins need not be regenerated. To regenerate the resin, first separate the cationic and the anionic particles by sedimentation in a large beaker of water. Then, thoroughly soak the lower (cationic) resin with 10% HCl and the upper (anionic) resin with 10% NaOH. Wash both with distilled water in separate Buchner funnels. The conductivity of the wash water from each resin should be approximately 0.20 μS (1 S = 1 ohm^{-1}). Remove all wash water, remix the cationic and anionic resin particles with fresh water in a beaker, and immediately collect the resin in a Buchner funnel. The final resin should be amber–green, indicating that the green anionic particles are well mixed with the cationic particles.

Dialysis of the Latex Particles

2. Place 2.0 mL of the 91-nm latex particles solution in a preboiled dialysis tube, seal the ends by tying off, and suspend the tubing in deionized water. Monitor the conductivity of the water over approximately 5 days, and change the water occasionally. When the conductivity of the water around the dialysis tube remains approximately equal to the conductivity of deionized water, withdraw the solution from the tubing and store it in a plastic vial (do not use glass!). The volume of this solution should be approximately 8 mL and the percent solids approximately 2.5%.

Preparation of Thin Cells

3. Assemble five thin cells consisting of two 2- × 2-inch square pieces of 0.25-inch-thick plexiglass separated by a 0.03-inch Teflon sheet with a 1- × 0.5-inch rectangle cut out of it (see Figure 3-4-1). Drill 6 holes in the assembly adjacent to the rectangular cut-out, and tighten the "sandwich" together with 6 machine screws, 6 nuts, and 6 washers. Make certain that the finished assembly holds water tightly.

Preparation of Latex Crystals

4. Add drops of the 2.5% latex solution to each of the five 1.0-cm^2 polystyrene cuvettes in the following amounts: 40 drops, 30 drops, 20 drops, 10 drops, and 5 drops. Use a pipette filled with ion-exchange resin to add water to the cuvettes to adjust the total volume of each to 40 drops (make certain that the effluent from the pipette tests at approximately 0.20 μS). To each of the five solutions, add 12 beads of low-conductivity resin from the same pipette to exchange residual ions in solution, and remix the solutions. The beads should settle quickly to the bottom of the cell.

5. From each of the cuvettes, withdraw liquid with a syringe and transfer it to one of the five thin cells to fill the void volume. Add 12 low-conductivity beads to the thin cell and seal the top with paraffin tape. Set the five thin cells and five cuvettes aside to crystallize.

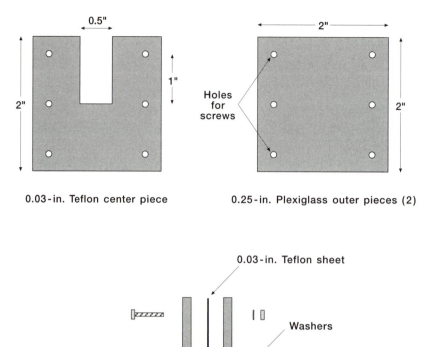

FIGURE 3-4-1. Construction of thin cells.

6. Within 10 days, crystals suitable for study will fill the containers. Avoid moving the samples as much as possible because the crystals are extremely fragile. If the crystals are accidentally damaged, however, recrystallization often takes place rapidly. The crystals can be stored for many weeks depending on the permeability of the container and the rate at which ions are leached from the container walls.

Diffraction from Latex Crystals

7. Figure 3-4-2 shows the experimental apparatus for examining diffraction patterns from the latex crystals in the thin cells. The thin cell is connected to an (x,y,z)-translator boom by means of a connector rod and suspended in a water bath to match approximately the indices of refraction of the latex, plexiglass, and surrounding medium, thus minimizing scattering and refraction of light.

8. Once the laser beam is properly positioned, attach a converging lens to the side of the water tank with small pieces of tape. We used a plano-convex lens with a 31-mm focal length. Shadows should be visible on a paper screen covering the outside

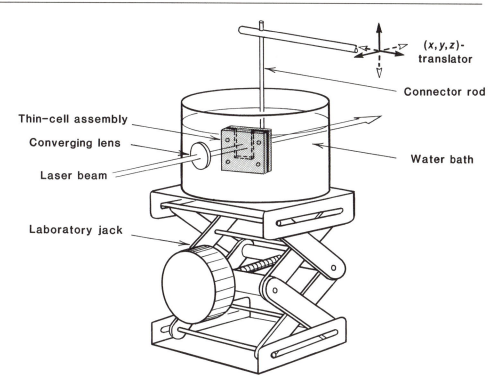

(x,y,z)- translator

Connector rod

Thin-cell assembly

Converging lens

Laser beam

Water bath

Laboratory jack

FIGURE 3-4-2. Schematic diagram of experimental setup for observing Kossel rings from colloidal crystals.

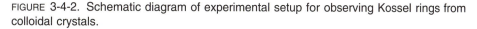

of the water bath. Use the *(x,y,z)*-translator to search for the cleanest pattern. Several samples may need to be tried to find a high-quality pattern.

9. Record the distance from the sample to the screen and the diameter of the central, dark ring. The Kossel rings also can be traced on tracing paper.

Forward-Scattered Kossel Patterns

10. If the colloidal crystals are of sufficient quality, a bright halo-shaped ring as well as a dark ring may be observed in the forward-scattering direction. If so, a water bath and crystallizing dish are not necessary. Such behavior is observed only for thin cells in which a single crystal spans the width of the cell.

Determination of the Bulk Modulus of a Single Crystal (Optional)

11. Select a sample cell that has a rather large single crystal, approximatey 1-cm high, with a single face presented to the viewer. Place the sample thin cell in a water bath, and focus the laser beam on a spot either at the top or bottom of the single crystal. Record the radius of the Kossel ring and the "*z*" position of the *(x,y,z)*-translator. Then raise or lower the sample holder so that the laser focuses on a different section of the same crystal. Record the diameter of the Kossel ring and the new *z* position of the *(x,y,z)*-translator.

12. Repeat this procedure four or five times until the other end of the crystal is reached. Plot the volume of the unit cell vs. depth in solution.

HAZARDS AND PRECAUTIONS

Never look into the laser beam.

DISPOSAL

No special precautions are necessary.

DISCUSSION

Charged polystyrene microspheres form milky colloidal suspensions in water, and the removal of contaminants by dialysis and ion-exchange treatment yields brilliant opal-colored solutions. Under conditions of low particle concentration and low ionic strength, the colloidal suspension can crystallize so as to minimize the Coulomb repulsions between the charged microspheres (2–4). (Please see the discussion section in experiment 2-4 for further discussion of interactions between latex microspheres.) The lattice spacings of these crystals are typically several times the diameters of the spheres, comparable to the wavelengths of visible light (2,5). When the beam of a red laser passes through a colloidal crystal, the dominant diffraction pattern consists of dark and light bands known as *Kossel rings* against a background of diffusely scattered light. Typical diffraction patterns have a central ring, frequently other overlapping peripheral rings, and occasionally a clean, fourfold symmetrical pattern.

Whereas Bragg diffraction spots arise from the diffraction of incident light from planes of scattering centers in a crystal, Kossel rings result from the diffraction of the light scattered inside the crystal. Thus, for Kossel rings to be prominent, the individual scattering centers in the crystal lattice must strongly scatter the incident light. Kossel rings are occasionally observed in X-ray diffraction studies of crystals, but most crystals do not effectively scatter X-rays, and the Kossel rings are often weak. Kossel rings are more commonly associated with the bombardment of metals with high-energy beams of electrons or protons (6). The electrons collide with the metal atoms in the lattice and scatter throughout the sample, generating Kossel ring diffraction patterns as described below.

The particles that make up colloidal crystals strongly scatter visible light, and Kossel rings are the dominant diffraction effect observed. Other researchers (7) have also observed Bragg diffraction spots, but in many studies, these spots are masked by the diffusely scattered light. To understand the origin of the Kossel rings, we can treat each of the latex spheres as a point source and consider how the crystal planes diffract the scattered light (8). The various planes of spheres in the crystal diffract the light scattered inside the crystal in exactly the same manner as they diffract the incident light, as in the formation of the Bragg diffraction spots. That is, the planes act as a partially reflecting, partially transmitting mirror.

Consider the diffraction of the light from one point source inside the crystal (point P) by one set of parallel planes of latex spheres. As depicted in Figure 3-4-3, each plane of spheres reflects some of the light, giving a transmitted beam T and a reflected beam R at each plane and for each angle of incidence. The reflected beams at most angles are not observed because the beams reflected from parallel planes interfere destructively.

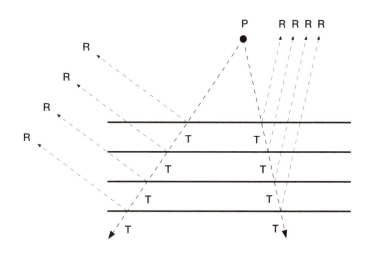

FIGURE 3-4-3. Examples of transmitted **(T)** and reflected **(R)** beams from parallel planes of latex spheres. Point **P** represents one point source scattering center (latex sphere) within the crystal lattice.

Only beams that reflect at the Bragg angle interfere constructively, which is determined by the Bragg condition

$$2\,d\,\sin\,\theta = n\,\lambda \tag{3-4-1}$$

Rearranging this expression gives

$$\theta = \sin^{-1}\left(\frac{\lambda}{2d}\right) \tag{3-4-2}$$

for first-order diffraction. The beams of light from the source that fulfill this condition form a cone that has a half-angle of

$$\phi = \frac{\pi}{2} - \theta \tag{3-4-3}$$

with an axis perpendicular to the reflecting planes (Figure 3-4-4). This cone of light is Bragg diffracted from the set of planes, resulting in a ring of slightly brighter light against the background of diffusely scattered light. A dark ring is observed in the opposite direction, which corresponds to the directions in which light was "borrowed" for the diffraction process.[1]

---

1. The theory we have just presented is somewhat of an oversimplification. One serious conceptual flaw is that in Figure 3-4-6, there will also be planes on the other side of the point source from which light will be Bragg diffracted. The planes on each side of point P create a bright ring and a dark ring, and the bright ring from one side coincides closely with the dark ring from the other side. Thus, the intensity borrowed from transmission by diffraction and vice versa should tend to cancel. As it turns out, if absorption (extinction) is also included in this treatment, then the cancellation is incomplete, and the Kossel rings appear as we had predicted (7,8).

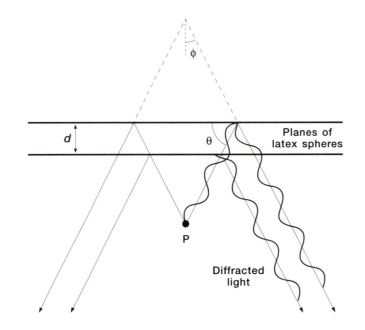

FIGURE 3-4-4. Origin of cones of constructive interference (Kossel rings). ϕ represents the half-angle of the cone, and θ is the Bragg angle, defined by $n\lambda = d\sin\theta$.

As seen in Figure 3-4-5, many planes of spheres surround each point source, and each set of planes can create a Kossel ring. In theory, thousands of Kossel rings could be observed from a single crystal (9), but in practice only a few are strong enough to be observed against the diffusely scattered light.

From equations 3-4-2 and 3-4-3, the size of a Kossel ring can be used to determine the spacing between the parallel planes of spheres that were responsible for the formation of the ring. Substituting equation 3-4-3 into equation 3-4-1 gives

$$2\,d\,\cos\,\phi = \lambda \tag{3-4-4}$$

Here, λ is the wavelength of the light within the colloidal crystal. This wavelength may be related to the vacuum wavelength λ_{vac} according to

$$\lambda = \frac{\lambda_{vac}}{n} \tag{3-4-5}$$

where n is the refractive index of the colloidal crystal. Hence,

$$d = \frac{\lambda_{vac}}{2n\,\cos\,\phi} \tag{3-4-6}$$

This expression allows for a rough estimate of the intersphere dimensions of the colloidal crystal. Ultimately, however, the goal of a diffraction experiment is to determine not

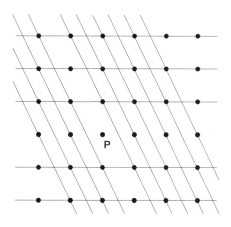

FIGURE 3-4-5. Two sets of parallel planes of spheres surrounding a point source **(P)**.

only how far apart the spheres are but also how they are arranged in three dimensions. The three most common crystal arrangements are simple cubic (SC), body-centered cubic (BCC), and face-centered cubic (FCC) (Figure 3-4-6). Equation 3-4-6 provides no help in distinguishing between these three cases, and determination of the three-dimensional arrangement of a crystal is generally difficult. Segschneider and Versmold (5) and Asher and Carlson (9) present outlines of how to distinguish between the SC, BCC, and FCC cases and to assign the Kossel rings to planes of spheres. With the use of low-power lasers, however, often only a few Kossel rings are visible, thereby making any assignment ambiguous.

Luckily, the Kossel ring diffraction patterns from latex sphere crystals have been studied extensively, and information presented in other studies allows us to assign the Kossel rings observed in this experiment. Pieranski (4) notes that, based on the Kossel rings observed in other studies, the SC arrangement can be ruled out as a possibility for the unit cell of the latex spheres under all conditions. Pieranski also generated theoretical

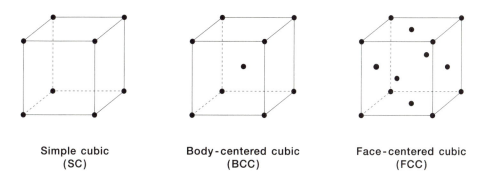

Simple cubic
(SC)

Body-centered cubic
(BCC)

Face-centered cubic
(FCC)

FIGURE 3-4-6. Three cubic crystalline packing arrangements. The BCC arrangement has one sphere in the center of the cube, and the FCC arrangement has one sphere in the center of each face.

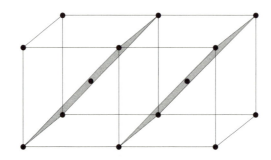

FIGURE 3-4-7. The planes of latex spheres known to be responsible for the dark inner ring observed in this experiment. The crystal structure is BCC, and the planes are denoted by the Miller indices of 110.

diffraction patterns for the FCC and BCC forms of the crystals at different concentrations of the latex spheres, and his findings indicate that the fourfold symmetrical pattern observed in this experiment is consistent with a BCC crystal. Further, the plane of spheres responsible for the dark inner ring observed in this experiment is denoted by the Miller indices (hkl) of 110 (Figure 3-4-7). (The Miller indices are a convenient way of denoting different crystal planes and are discussed in most introductory physical chemistry textbooks.) For a BCC crystal, the nearest neighbor interparticle separation distance d is related to the separation distance between planes, d_{hkl}, by (7)

$$d_{hkl} = \frac{2d}{\sqrt{3\ (h^2 + k^2 + l^2)}}$$
(3-4-7)

Based on a half-angle $\phi = 29.2°$ and a refractive index of $n = 1.34$, we found values of $d_{110} = 270$ nm and $d = 330$ nm for the crystal we studied. Using a simple geometric consideration of the BCC unit cell (2), we determined the length of the sides of the unit cell to be 380 nm.

Colloidal crystals display several interesting properties that can be studied with the apparatus used in this experiment. Monovoukas and Gast (10) observed diffraction patterns from latex crystals consistent with both the BCC and FCC configurations under different conditions. Figure 3-4-8 shows a phase diagram they generated to show conditions that favor the FCC or BCC crystalline configurations. An interesting follow-up experiment would be to modify the ionic strength or latex sphere concentration to attempt to prepare crystals of FCC or BCC configurations that have different Kossel ring patterns. In addition, colloidal crystals have a fairly well defined melting point, just as do crystals composed of atoms and molecules. Thus, another follow-up experiment would be to observe the changes in the Kossel ring patterns that occur near the melting point of the crystal. Finally, the pattern of Kossel rings observed sometimes depends on the wavelength of light used (7), and repeating the observations of the Kossel rings using different wavelengths of light might be rewarding.

Determination of the Bulk Modulus of a Single Crystal

Colloidal crystals are readily compressed, and the size of the unit cell in a single crystal in water decreases from the top of the crystal to the bottom. Because the size of

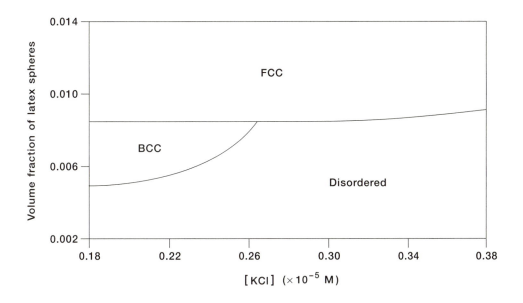

FIGURE 3-4-8. Phase diagram generated by Monovoukas and Gast (10), showing the conditions that favor FCC and BCC crystal structures.

the central Kossel ring directly reflects the size of the unit cell, the compression of the crystal can be monitored by observing the change in the size of the Kossel ring. The pressure in the solution is nearly proportional to the depth in solution, and the size of the central Kossel ring increases linearly with increasing column height. This relationship leads to a uniform decrease in the lattice spacing from top to bottom in a single crystal.

The isothermal compressibility β, defined as the negative of the change in volume per unit volume per change in pressure, and its reciprocal, the bulk modulus B, relate the volume change of the sample (in this case, the volume of the unit cell) to the pressure change in the surrounding medium according to the thermodynamic equation

$$B = -\left(\frac{\partial P}{\partial(\ln V)}\right)_T \qquad (3\text{-}4\text{-}8)$$

Asher and Carlson (9) describe how this formula can be used to derive an equation that relates the unit cell volume to the depth in solution, utilizing the chain rule for differentiation and an equation for the pressure vs. depth in a solution. The result of this derivation is

$$V - V_o = \frac{-4mg}{B}(z - z_o) \qquad (3\text{-}4\text{-}9)$$

where V (m^3) is the unit cell volume, m (kg) is the effective mass of the latex spheres in water (approximately 1.8×10^{-20} kg for a sphere 91 nm in diameter), g is the gravitational acceleration constant (ms^{-2}), B is the bulk modulus (N/m^2 = kg m^{-1}s^{-2}), and z is the vertical position (m) in solution. The variables z_o and V_o describe the unit cell volume and vertical position at some standard position. Thus, the slope of a plot of

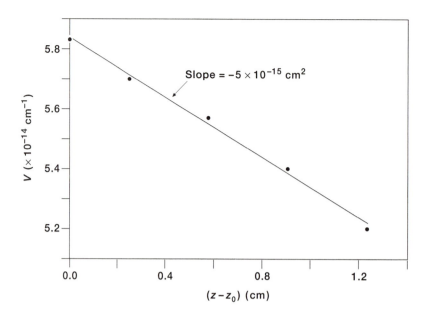

FIGURE 3-4-9. Plot of unit cell volume vs. depth of crystal in solution used to determine bulk modulus.

the volume of the unit cell vs. relative height (depth) of the crystal in solution (such as Figure 3-4-9) allows the determination of the bulk modulus (2,6,11). For latex crystals prepared from a solution of 91-nm spheres with a volume fraction of 0.0098, we estimated a bulk modulus of 1.4 N/m^2 or 14 dynes/cm^2. This value is much less than expected for a real crystalline solid.

An interesting variation is to prepare the colloidal crystals in heavy water (D_2O) rather than normal water (H_2O). Whereas the latex microspheres sink in regular water, they float in the heavy water (in equation 3-4-9, the value for the effective mass m would be negative). As a result, the lattice of the colloidal crystal in D_2O is compressed at the top of the column (11).

REFERENCES

1. American Association of Physics Teachers, Publications Department, 5112 Berwyn Rd., College Park, MD 20740.
2. Spencer, B., Zare, R.N. *J. Chem. Ed.* 68 (1991) 97.
3. Goodwin, J.W., Ottewill, R.H., Parentich, A. *J. Phys. Chem.* 84 (1980) 1580.
4. Pieranski, P., Dubois-Violette, E., Rothen, F., Strzelecki, L. *J. Physique* 42 (1981) 53; *J. Physique 41* (1980) 369.
5. Segschneider, C., Versmold, H. *J. Chem. Ed.* 67 (1990) 967.
6. Tixler, R., Wache, C. *J. Appl. Cryst.* 3 (1970) 466.
7. Cowley, J.M. *Diffraction Physics.* North-Holland, Amsterdam: 1981; Chapter 14.
8. Rundquist, P.A., Photinos, P., Jagannathan, S., Asher, S.A. *J. Chem. Phys.* 91 (1989) 4932.
9. Asher, S.A., Carlson, R.J. *Appl. Spectrosc.* 38 (1984) 297.
10. Monovoukas, Y., Gast, A.P. *J. Colloid. Int. Sci.* 128 (1989) 533.
11. Crandall, R.S., Williams, S. *Science* 198 (1977) 293.

CHAPTER 4

REFRACTION OF LIGHT

Light is a remarkable form of wavelike energy because it does not require a medium for transmission; in other words, it can travel through a vacuum. In a vacuum, light of different wavelengths and different frequencies travels at the same speed, c, whose approximate value is 2.99792×10^8 m/sec. What happens when light enters a medium? Because light is an electromagnetic wave, it interacts with the charged particles in the medium through which it is traveling, thereby creating an electromagnetic disturbance in the medium. As a result, the speed of light in the medium decreases by an amount dependent on the strength of this interaction. This slowing-down effect can be likened to swimming in molasses versus swimming in water; the former has the greater drag. The implication is that even though a material may be essentially transparent to light, the light may still interact with that material. Indeed, the less transparent the material, the more the light slows down in that material because of the greater interaction.

The relationship may be quantified by taking the ratio of c, the velocity of light in a vacuum (in m/sec), to v, the velocity of light in a medium (in the same units). This expression is called the refractive index:

$$n = c/v \tag{4-1-1}$$

where n is a dimensionless quantity. For a vacuum, the refractive index is unity, whereas the value of n in any medium is greater than unity. For example, n_{air} is very close to unity, n_{water} is 1.33, and n_{glass} is typically 1.5; the exact value depends on the type of glass. At 1 atm of helium, $n = 1.000036$, whereas for diamond, $n_{diamond} = 2.419$. In addition, the refractive index of a transparent material is weakly dependent on the wavelength of light. Red light travels slightly faster than blue light. The reason for this dependence is that at shorter wavelengths (higher energies), media such as water and glass begin to absorb the radiation. The more the medium absorbs the light, the more the interaction between the light and the medium increases. This variation in speed of light in a given medium as a function of wavelength is called *dispersion*. This phenomenon is exploited in a prism, for example, to separate white light into a rainbow of its constituent colors.

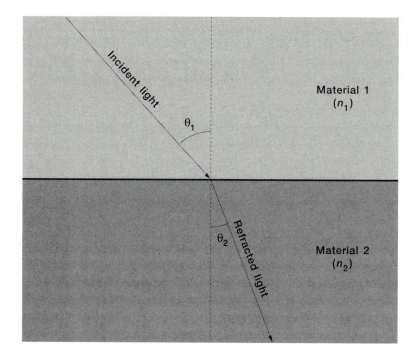

FIGURE 4-1. Refraction of a beam of light at the boundary between two materials.
In this diagram, $n_2 > n_1$. θ_1 is the angle of incidence, and θ_2 is the angle of refraction.

To understand how a prism works, we need to consider what happens when
a beam of light travels through two substances with different refractive indices,
such as glass and air. As Figure 4-1 shows, the beam of light bends (refracts) at
the boundary between the two substances. The angle at which the beam of light
strikes the boundary, measured from the surface normal (a line perpendicular to
the surface) is known as the *angle of incidence* (θ_1 in the figure), and the angle
at which it leaves the boundary is called the *angle of refraction* (θ_2). These two
angles are related by Snell's law:

$$n_1 \sin \theta_1 = n_2 \sin \theta_2 \qquad\qquad (4\text{-}1\text{-}2)$$

where n_1 and n_2 are the refractive indices of materials 1 and 2. A prism takes
advantage of dispersion, the change of the refractive index with wavelength, to
bend by different amounts the colors that make up white light.

Equation (4-1-2) tells us that if the refractive index of one substance is
known, the refractive index of another substance can be determined at some
wavelength by observing how a beam of light at that wavelength refracts at the
boundary between the two substances. Lasers are particularly useful for measuring refractive indices in this manner because laser light propagates so well as a
beam and is so pure in wavelength. Sophisticated methods of determining the
refractive index using lasers are discussed by Noll (1) and Edminston (2).

As a beam of light passes from a medium of smaller refractive index to one of larger refractive index, it is refracted more toward the normal of the medium with the larger refractive index. Thus, for $n_1 < n_2$, a critical angle of refraction $\theta_2^{critical}$ exists in medium 2 that cannot be reached by the light beam incident on medium 1 no matter what the angle of incidence. As the angle of incidence θ_1 approaches 90°, sin θ_1 approaches unity, and according to Snell's law, $\theta_2^{critical}$ is given by

$$\theta_2^{critical} = \sin^{-1}(n_1/n_2) \tag{4-1-3}$$

The inverse situation is also of great practical interest. If a light beam travels in the opposite direction, from a medium of a larger refractive index to one of a smaller index, the light beam is further bent away from the normal in the medium of the smaller refractive index. As the angle of incidence is increased, it reaches a value such that the angle of refraction becomes 90°; here, the refracted ray travels along the boundary between the two media. As the angle of incidence is increased further, Snell's law has no solution, and the light beam is reflected at the boundary rather than refracted. The critical angle for this behavior is called the *angle of total internal reflection*. This analysis explains why diamonds sparkle. (Recall that a diamond has a very large refractive index.)

Total internal reflection also allows us to construct waveguides for light, such as a glass fiber that traps and channels light inside it by repeated total internal reflection at the walls of the fiber. The above considerations also explain why a glass optical fiber continues to work even when placed in water. Cladding the glass fiber to produce a graded refractive index profile greatly reduces the dispersion in the fiber so that a pulsed light signal can traverse the fiber with small distortion in time. Consequently, low-loss graded index fibers that carry laser-generated signals are becoming widely used in the communications industry.

This chapter presents two applications in which a low-power red laser is employed to detect changes in the refractive index. Either a red He–Ne laser (632 nm) or a red diode laser (670 nm) can be used, but switching between these two laser sources reveals a small difference because the two wavelengths are refracted at different angles. The first application is a laser-based detection system for column chromatography. Traditionally, the application of column chromatography in educational settings has been limited to colored compounds, whose separation can be detected visually. Experiment 4-1 describes how a laser can be used to detect this separation when colorless compounds elute from a column; the detection is based on small changes in the refractive index of the eluent. This detection system is simple to assemble and requires little equipment: a laser, a microscope objective, and a light detector, such as a photometer or a photodiode. Experiment 4-1 also describes how to construct and operate a simple chromatography column using a buret. Experiments 4-2 and 4-3 provide two sample applications of the system.

The second application is the use of a laser to detect refractive index changes during the course of a chemical reaction. The reaction studied is the acid-catalyzed conversion of glycidol to glycerol, in which the product has a higher refractive index than the reactant. The measurement of the refractive index of the reaction mixture as a function of time yields kinetic information, and performing the reaction at several temperatures permits determination of the activation energy of the reaction.

REFERENCES

1. Noll, E.D. *Phys. Teach.* 11 (1973) 307.
2. Edminston, M.D. *Phys. Teach.* 24 (1986) 160.

COLUMN CHROMATOGRAPHY WITH A LASER-BASED REFRACTIVE INDEX DETECTION SYSTEM

This experiment explains how to construct and use a chromatography column with a laser-based detection system.

DEGREE OF DIFFICULTY

Experimental: moderate
Conceptual: moderate

MATERIALS

- laser
- light-detection system
- buret
- solvent reservoir (if desired)
- heptane (or other desired mobile phase) (Sigma product #H9629. Approx. $18 for 1 liter)
- detection cell (constructed of glass tubing, see Figure 4-1-1); the diameter and wall thickness of the glass tubing are not critical
- microscope objective (or other similar lens)
- glass wool
- purified sand
- powder funnel
- syringe (such as an insulin syringe)
- silica gel (or other desired stationary phase)
- ring stand and clamps
- 30 mL of methanol
- various organic compounds, such as naphthalene, o-dichlorobenzene, and benzoic acid

PROCEDURE

Loading the Column

1. Add 50 mL of heptane (mobile phase) to the column with the stopcock closed. Next, add a plug of glass wool and then a small amount of purified sand to the bottom of the column. If any sand adheres to the sides of the column, wash it down with a small amount of heptane, and then allow a few milliliters of heptane to drain from the column.

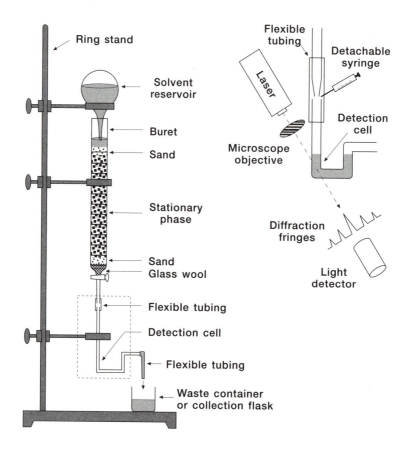

FIGURE 4-1-1. Schematic representation of column chromatography system and laser-based refractive-index-detection system.

2. Using a powder funnel, pour silica gel (stationary phase) into the column so that it gently settles in the heptane without clumping. Continue adding the stationary phase until it fills approximately two-thirds of the column; make certain that the liquid level stays above the gel at all times. Drying of the column leads to cracking, which results in poor column performance. Help the stationary phase settle by slowly draining approximately 30 mL of heptane from the column.

Setting Up the Detection System

3. Create a detection cell composed of glass tubing with a shape similar to that depicted in Figure 4-1-1. Attach the detection cell to the tip of the buret using flexible tubing and allow it to fill with mobile phase. Direct the laser beam to graze the detection cell. The laser beam should pass through nonturbulent liquid in the detection cell (Figures 4-1-1 and 4-1-2). Focus the laser beam with a microscope objective to produce a diffraction pattern that consists of a series of bands of decreasing light inten-

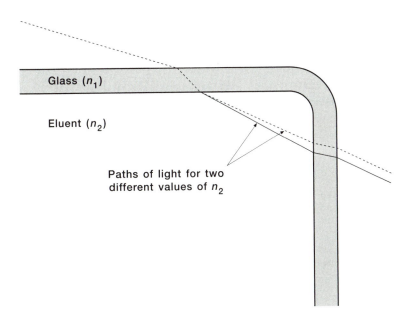

FIGURE 4-1-2. Migration of laser beam through detection cell. The path changes as a result of changes in the refractive index of the eluent.

sity. While solvent flows through the detection cell, center the light detector on one of the most clearly resolved light zones.

4. Dissolve a small amount of an organic substance (such as those suggested in the list of materials) in heptane and fill a syringe with this liquid. Insert the syringe needle into the flexible tubing that connects the buret to the detection cell, and inject enough of the solution to expel the heptane already in the detection cell. The introduction of the solute to the mobile phase should cause a deflection in the laser beam and a decrease in the light intensity measured by the detector. If it does not, try readjusting the detection system or using a higher concentration of solute. If the injected solution causes multiple fringes to move past the photodiode, lower the concentration of the solute. The detection limits of the apparatus are addressed in the discussion section.

Loading and Elution of Solutes

5. Once the detection system is clearly working properly, prepare a 2:1 mixture, grain by grain, with a total mass of approximately 10 mg dissolved in a minimal amount of heptane.

6. Lower the level of the mobile phase to just above the top of the gel. Add the mixture to the top of the column, and gently wash the sample into the column with two small portions of mobile phase using a squirt bottle or disposable pipette. After each washing, lower the level of the mobile phase to slightly above the top of the gel. After

the two washings, add a band of sand to the top of the column, so that the addition of more mobile phase disturbs the column as little as possible. Keep the level of the sand under the level of the liquid.

7. Fill the remaining portion of the tube with mobile phase. If desired, fill a reservoir container with mobile phase, and place it at the top of the column as shown in Figure 4-1-1. If the reservoir is not used, mobile phase will need to be added to the column during the separation.

8. Allow the mobile phase to elute from the column at approximately 1 to 2 mL per minute. To minimize the escape of mobile phase vapors, have the eluent flow directly into a jug from the detector cell through a piece of flexible tubing. When eluent that contains a solute flows through the detection cell, the laser beam will be deflected, resulting in a decrease in the intensity of light measured. A photometer can be used in conjunction with a chart recorder to provide a continuous trace of light intensity over time. The components of the mixture can be collected as they elute from the column by changing the collecting flask each time a decrease in light intensity is observed. At the end of a run, wash the column with methanol and then rewash it with heptane to remove all traces of methanol.

HAZARDS AND PRECAUTIONS

Heptane is flammable. Avoid breathing heptane fumes. Naphthalene, o-dichlorobenzene, and benzoic acid are combustibles. Keep them away from flames. o-Dichlorobenzene and methanol are skin-absorbable poisons, so wear gloves when handling them. Wear eye protection at all times.

DISPOSAL

Dispose of the eluent as organic waste. Naphthalene and o-dichlorobenzene may require special disposal procedures.

DISCUSSION

Chromatography is a method for separating chemicals. The chromatography system described in this experiment is classified as liquid-solid chromatography because the key components of the system are a solid packing, which is known as the *stationary phase,* and a liquid that flows through it, which is called the *mobile phase.* A mixture to be separated is introduced at the top of the column (a buret in this experiment), and the components of the mixture are allowed to flow through the column with the mobile phase. The components of the mixture have different degrees of attraction for the stationary phase, and molecules in the mixture alternately attach, detach, and reattach to the stationary phase. Those components of the mixture that have a greater attraction for the stationary phase spend a greater amount of time adhered to the stationary phase and thus flow more slowly through the column. Ideally, each of the components of a mixture travels through the column at a different rate and elutes (drains out) from the column at a different time. The aliquots containing the various components of the mixture can be collected and evaporated, if desired, to recover the components of the initial mixture.

The mechanisms that govern the interactions between the stationary phase and the components of the mixture vary from application to application. The most common stationary phases are polar materials, such as alumina (Al_2O_3), powdered sugar, or silica gel, which is used in this experiment. Polar solutes adhere to polar stationary phases more strongly, because of the dipole–dipole, ion–ion, ion–dipole, and hydrogen-bonding interactions that exist between polar materials. Thus, polar compounds travel more slowly through such columns.

The choice of mobile phase affects how long the various components remain in the column. If a nonpolar mobile phase is employed, the nonpolar solutes elute quickly, but the polar compounds adhere strongly to the packing and elute slowly. Employing a more polar solvent as the mobile phase introduces greater competition between the solvent and the stationary phase for polar substances, and polar compounds elute more quickly in a more polar solvent than they do in a less polar one. Table 4-1-1 lists several solvents commonly used as mobile phases in order of increasing polarity. The combina-

TABLE 4-1-1. SOME SOLVENTS COMMONLY USED FOR COLUMN CHROMATOGRAPHY

Relative Polarity	Solvent
Increasing Polarity ↓	Heptane
	Toluene
	Ethanol
	Methanol
	Water

tion of silica gel as the stationary phase and heptane as the mobile phase is useful for many separations of organic compounds, but optimizing the success of a separation by experimenting with other mobile and stationary phases is often possible. Mobile phases need not be pure solvents; mixtures of two miscible solvents may achieve better results.

The term *chromatography* takes its meaning from its early applications, which involved the description ("graphy") of separation of colored substances ("chromato"). The column chromatography system described in this experiment can be used to separate colored compounds, in which case bands of color traveling down the column can be observed and collected as they elute from the column. The vast majority of organic compounds are colorless in solution, however, and some other method must be devised for detecting when bands of solute elute from the column. Many such detection systems have been devised, including those based on ultraviolet spectrophotometry, polarography, conductivity, mass spectroscopy, optical rotation, and light scattering (1), but the equipment necessary for these techniques is generally expensive.

The detection system described in this experiment relies on small changes in the refractive index of the eluent from the column. The light beam from a laser passes through air, the glass walls of the detection cell, and the eluent from the chromatography column, all of which have different indices of refraction. The refractive indices of air and the glass are constant at constant temperature, but the refractive index of the eluent depends on whether the eluent is pure mobile phase or a solute is dissolved in it. The varying refractive index of the eluent causes the laser beam to migrate slightly in the way shown in Figure 4-1-2. Inserting a microscope objective into the path of the laser beam results in a diffraction pattern from the objective's glass walls, and the migration of the laser beam can be detected by centering the light detector on one of the diffraction spots and noting the change in intensity as the beam migrates.

The laser-based system can detect variations of refractive indices on the order of parts per thousand. On the other hand, if the concentration of some component of the mixture in the eluent is too high, more than one diffraction fringe may pass by the light detector, resulting in a multiplet of peaks. As a rule of thumb, the technique works best if the solution containing the mixture to be separated has a concentration of approximately 0.1 to 20 mg per mL of mobile phase if the retention time is short and 0.5 to 40 mg per mL if the retention time is long. Substances retained longer are often peak broadened and hence less concentrated as they elute. To accommodate larger quantities of analyte, position the light detector on one of the less well resolved diffraction spots, where the sensitivity of the light intensity to refractive index change is greatly reduced.

Although the laser-based detection system does not aid in identifying the compounds eluting from the column, guessing which peaks correspond to which components of a mixture is often possible if the polarities of the compounds are known. A more rigorous identification procedure is to add the pure substances individually to the column and compare the elution times to the components in the sample. Quantitation of the peaks is difficult because the response of the detector is not necessarily linear for any given compound. In principle, a calibration curve could be prepared for each compound.

Figure 4-1-3 depicts the structures of some compounds that can be separated using the column chromatography system described in this experiment. A mixture of benzoic

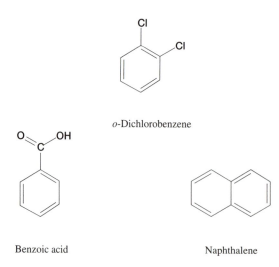

o-Dichlorobenzene

Benzoic acid

Naphthalene

FIGURE 4-1-3. Structures of compounds suggested for testing the column chromatography system.

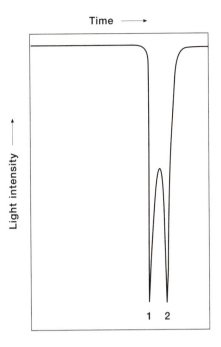

FIGURE 4-1-4. Chromatogram depicting the incomplete separation of 40 mg of *o*-dichlorobenzene and 50 mg of *p*-dichlorobenzene. The mobile phase is hexane, and the stationary phase is silica gel.

acid and naphthalene can be completely separated because the polarities of the two compounds are vastly different. On the other hand, we found that *o*-dichlorobenzene and *p*-dichlorobenzene could not be separated completely (Figure 4-1-4), probably because they are chemically very similar. The use of different mobile or stationary phases may allow a more successful separation of these two compounds.

REFERENCE

1. Vickrey, T.M., ed. *Liquid Chromatography Detectors.* Marcel Dekker; New York: 1983.

EXPERIMENT 4-2

EXTRACTION AND PURIFICATION OF LIMONENE

Limonene is extracted from orange peels and purified using the column chromatography system described in experiment 4-1. The laser-based detection system helps determine when the limonene elutes from the column.

DEGREE OF DIFFICULTY

Experimental: moderate
Conceptual: moderate

MATERIALS

- chromatography column with laser-based detection system described in experiment 4-1
- 2 oranges
- large mortar and pestle
- 200 mL of heptane (Sigma product #H9629. Approx. $18 for 1 liter)
- 10 g of anhydrous magnesium sulfate
- 125-mL separatory funnel
- Erlenmeyer flasks, beakers
- wide-mouth funnel and filter paper
- 18 live flies and 3 small jars *(optional)*

PROCEDURE

Extraction of Limonene

1. Grind the skins of two oranges with a mortar in a large pestle, adding heptane as needed to yield a mixture with the consistency of mush. Drain the orange organic liquid, and gravity filter it using the funnel and filter paper. Pour the orange liquid into a separatory funnel. Add approximately 20 mL of water, shake well, and drain the water out of the separatory funnel. Repeat this washing process, and pour the organic liquid into a beaker containing anhydrous magnesium sulfate to remove any remaining water. Pour off the organic liquid into another beaker, and allow the heptane to evaporate overnight, or speed the process by gently heating the mixture with a boiling stone. Keep in mind that heptane is flammable, and the safest way to heat heptane is in a sand bath on a heating pad. Do not use flames! Once the heptane has completely evaporated, there should be approximately 200 mg remaining of a light oil that smells like oranges.

Purification of Limonene

2. Assemble a chromatography column with a laser-based detection system as described in experiment 4-1. Dissolve approximately 50 mg of the oil in a minimal amount of heptane, and apply approximately half of this solution to a silica gel column as described in experiment 4-1. Elute the sample at a rate of approximately 1 mL/min.

The limonene fraction corresponds to the largest peak. Save this fraction and evaporate the solvent to yield a slightly colored oil.

Bioactivity of Limonene (Optional)

3. Place each of the following in a separate screw-top jar and observe periodically:
   a. Six live flies
   b. Six live flies and a few small pieces of orange rind
   c. Six live flies and a drop of purified limonene spread on a tissue

HAZARDS AND PRECAUTIONS

Heptane and limonene are flammable. Use a sand bath on a hot plate to heat heptane solutions, and work under a fume hood. Always use boiling stones when heating solutions. Avoid breathing heptane fumes. Wear eye protection at all times.

DISPOSAL

Dispose of the eluent as organic waste.

DISCUSSION

This experiment illustrates the usefulness of solid–liquid chromatography in isolating organic compounds from a complex mixture, such as orange peels. Many organic compounds that are important to the processed food and pharmaceutical industries are isolated from natural substances in a conceptually similar manner.

Limonene is a natural insecticide found in the peels of many citrus fruits. Its structure is depicted in Figure 4-2-1. Many types of insects become paralyzed and then die

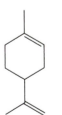

FIGURE 4-2-1. Molecular structure of limonene.

when exposed to limonene (1,2). In nature, limonene may help protect citrus fruits from fruit fly larvae, and currently limonene is being used commercially as an antiflea agent for pets.

REFERENCES

1. Beatty, J.H. *J. Chem. Ed.* 63 (1986) 768.
2. Coats, J.R., Karr, L.L., Drewes, C.D. *Naturally Occurring Pest Bioregulators,* ACS Symp. Series 449, American Chemical Society, Washington, DC: 1991; 305–316.

SYNTHESIS AND PURIFICATION OF OIL OF WINTERGREEN

Oil of wintergreen is synthesized from salicylic acid and purified using the column chromatography system described in experiment 4-1. The laser-based detection system helps determine when the oil of wintergreen elutes from the column.

DEGREE OF DIFFICULTY

Experimental: moderate
Conceptual: moderate

MATERIALS

- chromatography column with laser-based detection system described in experiment 4-1
- 0.5 g of salicylic acid (Sigma product #S0875. Approx. $6 for 100 g)
- 0.5 g of anhydrous magnesium sulfate
- 5 mL of methanol
- concentrated sulfuric acid
- mobile phase: heptane or 50% heptane/50% methylene chloride
- 25 mL of 10% sodium bicarbonate
- 25 mL of ether
- 125-mL separatory funnel
- Erlenmeyer flasks
- Bunsen burner or hot plate
- test tube
- glass funnel and filter paper
- boiling stone

PROCEDURE

Preparation of Oil of Wintergreen

1. In a test tube, mix 0.5 g of salicylic acid powder and 1.0 mL of methanol. Add 5 drops of 18-M sulfuric acid and a boiling stone. Under a fume hood, heat the mixture to a gentle boil on a hot plate, and allow the solution to boil for 5 minutes. Keep the volume of liquid steady by adding methanol as needed.

Prepurification of Oil of Wintergreen

2. Allow the reaction mixture to cool, and then pour it into 25 mL of water in a separatory funnel. Add 25 mL of ether and shake, venting frequently. Drain off the water and extract the ether layer with 25 mL of 10% sodium bicarbonate. Drain the aqueous layer, wash once more with water, and pour the ether layer onto magnesium sulfate in an Erlenmeyer flask. Allow the flask to sit for 10 minutes; then filter the

liquid through filter paper into a separate Erlenmeyer flask. Concentrate the liquid to an oil by evaporation or gentle heating. Recall that ether is highly flammable; use a sand bath to heat the solution and use a boiling stone. Do not use flames!

Column Chromatography of Oil of Wintergreen

3. Construct the chromatography column with laser-based detection system as described in experiment 4-1. Pure heptane can be used as the mobile phase, but a 50% heptane/50% methylene chloride mixture gives better results. Other mobile phases of similar polarity could be substituted for the heptane–methylene chloride mixture.

4. Dissolve the oil in a small amount of the mobile phase, and apply it to the column as described in experiment 4-1. The major peak in the chromatogram corresponds to oil of wintergreen. Allow the heptane to evaporate from the oil of wintergreen fraction by leaving the solution in a beaker under a fume hood for at least 24 hours. Alternatively, the heptane solution can be gently heated with a boiling stone. Keep in mind that heptane is flammable, and the safest way to heat heptane is in a sand bath on a heating pad. Do not use flames!

HAZARDS AND PRECAUTIONS

Methylene chloride is a suspected carcinogen. Methanol is a poison and can be absorbed through the skin. Heptane and ether are flammable; keep them away from flames. Avoid breathing heptane fumes. Wear eye protection at all times. Wear gloves when handling any solution containing methanol.

DISPOSAL

Methylene chloride, along with all halogenated solvents, requires special disposal procedures.

DISCUSSION

Oil of wintergreen has a pleasant odor and for this reason is used in such commercial products as cleaning solutions and soaps. The chemical structure of oil of wintergreen is similar to that of salicylic acid, and Figure 4-3-1 shows how oil of wintergreen is synthesized from salicylic acid in this experiment. Note that a derivative of salicylic

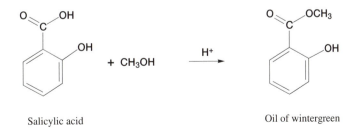

FIGURE 4-3-1. Acid-catalyzed reaction of salicylic acid with methanol to form oil of wintergreen.

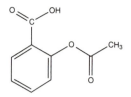

FIGURE 4-3-2. Molecular structure of acetylsalicylic acid.

acid, acetylsalicylic acid, is the active ingredient of aspirin (Figure 4-3-2). The synthesis of oil of wintergreen performed in this experiment is an example of a Fischer esterification reaction, which is described in most introductory organic chemistry textbooks. The accepted mechanism of this reaction is depicted in Figure 4-3-3.

The reaction mixture prepared in step 1 contains oil of wintergreen, unreacted salicylic acid, and other impurities. To purify the oil of wintergreen, the reaction mixture is first extracted with aqueous sodium bicarbonate solution, which removes most of the salicylic acid and other water-soluble impurities. The resulting mixture is purified further using column chromatography. The laser-based detection system helps determine when oil of wintergreen elutes from the column.

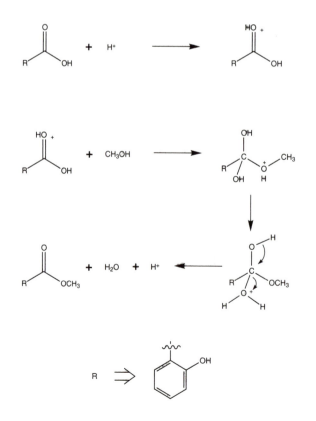

FIGURE 4-3-3. Mechanism of the reaction depicted in Figure 4-3-1.

This experiment illustrates the usefulness of column chromatography for separating and purifying the products of organic reactions. Very few organic reactions go to completion or yield only one product, and an important task for a synthetic organic chemist is to separate and identify the products of a reaction. The technique of column chromatography employed in this experiment could be applied to many other organic reactions, although the choice of mobile and stationary phases depends on the substances to be separated.

KINETICS OF THE ACID-CATALYZED CONVERSION OF GLYCIDOL TO GLYCEROL

This experiment demonstrates how laser-based measurements of the refractive index of a reaction mixture can yield kinetic information about a chemical reaction, specifically, the acid-catalyzed conversion of glycidol to glycerol.

DEGREE OF DIFFICULTY

Experimental: moderate
Conceptual: moderate to difficult

MATERIALS

- red laser
- test tube (20-mm inner diameter, 50-mL volume)
- 6.0 mL of glycidol (Sigma product #G0887. Approx. $21 for 100 mL. $d = 1.12$ g/mL. MW = 74.08 g/mol)
- 10 mL of concentrated perchloric acid ($\sim 60\%$ $HClO_4$ by weight)
- rubber stopper with one hole for a thermometer
- thermometer
- disposable pipette
- ring stand and clamps
- water bath and ice chips
- magnetic stirrer and stir bar

PROCEDURE

1. In a fume hood, add 6.0 mL of glycidol, 14.0 mL of deionized water, and a stir bar to the test tube. Cap the test tube with a rubber stopper through which a thermometer is inserted.

2. Clamp the test tube over a magnetic stirrer located approximately 2 m from a screen (Figure 4-4-1). Direct the laser so that the beam travels across the diameter of the test tube, and then slowly move the test tube perpendicularly to the laser beam until the angle θ (Figure 4-4-2) reaches a maximum value, at which point the reflected image becomes noticeably sharper. Fix the test tube in this position.

Study of Reaction Kinetics

3. Using a disposable pipette, add 2 drops of perchloric acid to the test tube to initiate the reaction. Monitor the beam displacement as a function of time.

4. Repeat steps 1–3, but use different amounts of perchloric acid. We performed trials with 4, 6, 8, and 10 drops of acid. For 8–10 drops of acid, the reaction is rapid at room temperature and evolves a significant amount of heat; thus, it necessitates the use of a water bath and ice chips to regulate the temperature.

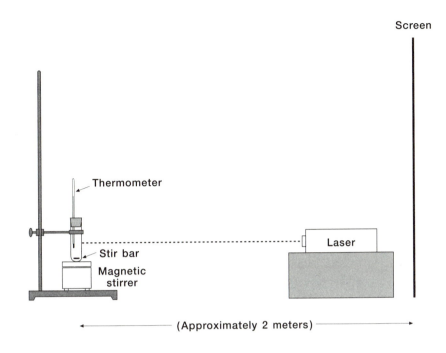

FIGURE 4-4-1. Schematic diagram of experimental setup for monitoring the refractive index of a reaction mixture to obtain kinetic data.

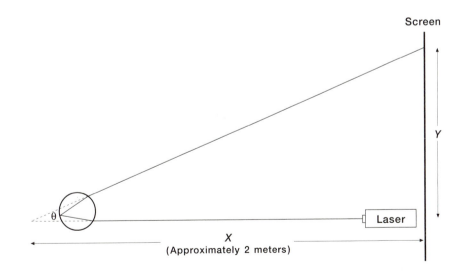

FIGURE 4-4-2. Definition of the distances X and Y and the angle θ.

Determination of Activation Energy (Optional)

5. Repeat the experiment at different temperatures using a constant concentration of perchloric acid and glycidol. Employ a water bath to maintain as constant a temperature as possible. Regulation of temperature is somewhat tricky because of the exothermicity of the reaction but can be accomplished in the well-stirred solution by judicious addition of ice while monitoring the temperature constantly. **Do not lean over the opening of the test tube!** See also a comment on temperature control in the discussion section under "Possible Modifications of Procedure."

HAZARDS AND PRECAUTIONS

Glycidol, as an organic epoxide, is a suspected chemical carcinogen. Glycidol and perchloric acid are both corrosive. Perchloric acid is a strong oxidizer and can cause chemical burns. Glycidol is combustible and requires refrigeration. The reaction can be safely monitored in a large test tube not more than 33% filled with up to 30% (volume) of glycidol in water, beyond which concentration heating effects become significant. Do not perform experiments at higher concentrations! A reaction that we attempted with 60% (volume) glycidol produced dangerous splattering.

DISPOSAL

Neutralize the reaction mixture and dispose of it as organic waste.

DISCUSSION

Glycidol belongs to a class of compounds known as epoxides. Epoxides have in common a three-membered ring that consists of two carbon atoms and one oxygen atom. Compounds containing three-membered rings are rarely highly stable, because the geometry of the ring forces the bond angles between the atoms to be near 60°, much smaller than the 109.5° bond angles normally associated with carbon atoms in sp^3 hybridization states. The decreased stability caused by undesirable bond angles is frequently referred to as *ring strain.*

The ring strain associated with epoxides causes them to be highly reactive, and epoxides generally react in ways that open up the three-membered ring (2). In this experiment, glycidol reacts with water under acid catalysis to form glycerol (Figure 4-4-3). The index of refraction of the product is greater than that of the reactant, and as the reaction progresses, the index of refraction of the reaction mixture increases. A laser

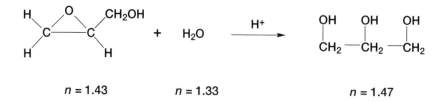

FIGURE 4-4-3. Acid-catalyzed ring opening of glycidol with water.

is used to monitor the index of refraction of the reaction mixture over time, thereby permitting quantitative measurements to be made of rate constants.

How Refractive Index Changes Reflect the Extent of a Reaction

In general, the index of refraction of a mixture (n) is a weighted average of the indices of refraction of the components (n_1, n_2, etc.)

$$n = n_1 X_1 + n_2 X_2 + n_3 X_3 + ... \qquad (4\text{-}4\text{-}1)$$

where X_1, X_2, etc. are the mole fractions of the various components. The reaction studied in this experiment is a three-component system consisting of glycidol, water, and glycerol. For the sake of concreteness, we can name glycidol as compound 1, water as compound 2, and glycerol as compound 3. Because

$$X_1 + X_2 + X_3 = 1 \qquad (4\text{-}4\text{-}2)$$

equation 4-4-1 can be rewritten as

$$n = n_1(1 - X_2 - X_3) + n_2 X_2 + n_3 X_3 \qquad (4\text{-}4\text{-}3)$$

which can be rearranged to

$$n = n_1 + (n_2 - n_1)X_2 + (n_3 - n_1)X_3 \qquad (4\text{-}4\text{-}4)$$

Filling in numerical values for the indices of refraction gives

$$n = 1.43 - 0.1(X_2) + 0.04(X_3) \qquad (4\text{-}4\text{-}5)$$

Under the reaction conditions employed in this experiment, the value of X_2 remains approximately constant during the course of a reaction, and the first two terms in equation 4-4-5 can be considered constant. The value of X_3 is a measure of the extent of the reaction and ranges from 0 at the outset of the reaction to approximately 0.12 if the reaction goes to completion. Thus, n would be expected to increase linearly with the extent of the reaction. For a more rigorous discussion of the linearity of the refractive index of the mixture versus the extent of the reaction, please see Appendix 4-4-A1 (after this experiment).

Laser-Based Monitoring of the Refractive Index of the Reaction Mixture

The system used to monitor the refractive index is based on an experiment described by Noll (3) that uses a laser to determine the index of refraction of a liquid in a test tube. The usefulness of this technique is that the displacement of the reflected laser beam (Y in Figure 4-4-2) is directly proportional to the refractive index of the solution. The mathematical argument presented by Noll to confirm this fact is somewhat complicated and is summarized in Appendix 4-4-A2. Understanding this argument, however, is not necessary to understand the results of this experiment. Figure 4-4-4 provides an empirical demonstration that the displacement Y is directly proportional to the refractive index by presenting a plot of experimental values of the beam displacement for various solutions of glycerol in water that have indices of refraction that range from 1.33 to 1.39.

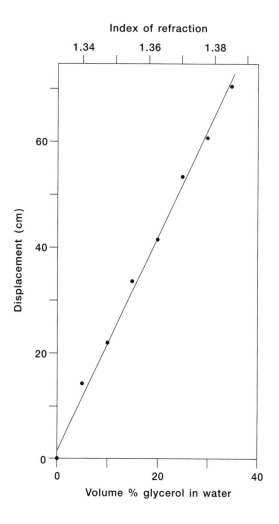

FIGURE 4-4-4. Empirical demonstration that beam displacement is directly proportional to refractive index using solutions of glycerol in water. The beam displacement is referenced to the displacement for pure water.

Experimental Results and the Reaction Mechanism

Figure 4-4-5 depicts the natural logarithm of the displacement (Y) versus time for different amounts of perchloric acid. The logarithm of Y over time is linear for each of the trials, indicating that the reaction is first order with respect to glycidol. This conclusion has been confirmed by two other studies of this reaction (4,5) using a different technique. The rate constants (k) are given by the slopes of the lines.

The accepted mechanism for the conversion of glycidol to glycerol[1] (2) (Figure 4-4-6) helps explain these results. The rate-limiting step in this mechanism is the second

---

1. Note that addition of water to the molecule occurs at the more highly substituted carbon. This fact has been ascertained by isotopic studies of the reaction (2,5) and initially caused confusion over whether the addition of water followed a concerted reaction or involved two steps—the ring opening of glycidol followed by the addition of water to the resulting carbocation. The current explanation (2) is that the addition of water is *primarily* a carbocation reaction, although the addition occurs by an S_N2 mechanism. The addition occurs at the more highly substituted carbon because the positive charge is more stable there.

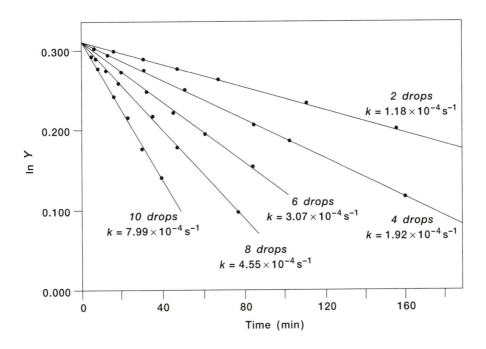

FIGURE 4-4-5. Kinetic data for the acid-catalyzed glycidol–glycerol reaction for various concentrations of perchloric acid.

Step 1: protonation of glycidol

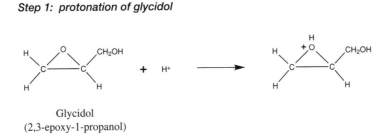

Step 2: nucleophilic attack of water on protonated glycidol

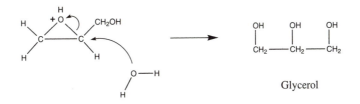

FIGURE 4-4-6. Accepted mechanism for the acid-catalyzed conversion of glycidol to glycerol.

step, the ring opening of protonated glycidol. This step involves two chemical species, and its rate is given by

$$\text{rate} = k_2[\text{water}][\text{protonated glycidol}] \tag{4-4-6}$$

where k_2 is a rate constant. If the rate of the first step is assumed to be so much faster than that of the second step that the first step can be considered to be at equilibrium, then the concentration of protonated glycidol should depend on both the concentration of glycidol and the acid concentration. This fact can be demonstrated by writing the equilibrium expression for the first step as

$$K = \frac{[\text{protonated glycidol}]}{[\text{H}^+][\text{glycidol}]} \tag{4-4-7}$$

(K is the equilibrium constant) and substituting this expression into equation 4-4-6:

$$\text{rate} = k_2 K[\text{water}][\text{H}^+][\text{glycidol}] \tag{4-4-8}$$

Because the concentration of water is very nearly constant, equation 4-4-8 can be simplified to

$$\text{rate} = C[\text{H}^+][\text{glycidol}] \tag{4-4-9}$$

where $C = k_2 K [\text{water}]$. Because the acid is a catalyst, the concentration of H^+ also does not change significantly during the course of the reaction. Thus, to a good approximation, the reaction appears to be a first-order reaction (Figure 4-4-5). Because the reaction is not rigorously first order, however, the rate constants are properly called pseudo–first-order rate constants.

This mechanism also predicts that the pseudo–first-order rate constants should increase linearly with acid concentration. Figure 4-4-7 shows that at lower acid concentrations, the rate constants increase approximately linearly with acid concentration, but that significant deviations from linear behavior are observed at higher acid concentrations. The fundamental explanation for this behavior is that the activity rather than the concentration of the hydrogen ion determines the rate of reaction. In the discussion of the mechanism of the reaction above, we employed the usual assumption that the activity of the various species could be closely approximated by the concentrations of the species. As the ionic strength (electrolyte concentration) of the reaction mixture increases, however, this approximation becomes less valid. Appendix 4-4-A3 provides a more detailed discussion of the dependence of the rate constants on acid concentration.

Determination of the Activation Energy

The activation energy of the reaction can be determined by measuring the rate constant of the reaction at different temperatures. The Arrhenius law states that the rate constant $k(\text{s}^{-1})$ of a reaction changes with temperature $T(°\text{K})$ according to the equation

$$k = A \exp\left(-\frac{E_a}{RT}\right) \tag{4-4-10}$$

where A is a frequency factor (s^{-1}), E_a is the activation energy (J mol^{-1}), and R is the

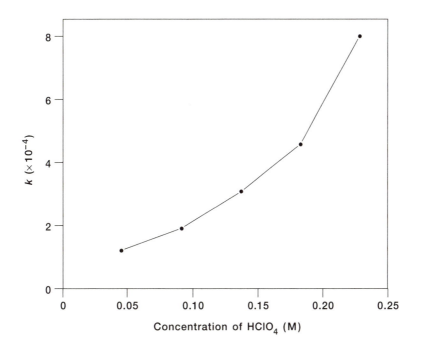

FIGURE 4-4-7. Dependence of pseudo–first-order rate constant (k) on perchloric acid concentration.

gas constant, 8.314 J $°K^{-1}$ mol^{-1}. Taking the natural logarithm of both sides gives

$$\ln k = \ln A - \frac{E_a}{RT} \tag{4-4-11}$$

Thus, a plot of $\ln k$ vs. $1/T$ allows the determination of the activation energy from the slope of the regression line. Figure 4-4-8a shows the displacement vs. time for the reaction catalyzed by 0.137-M (6 drops) perchloric acid at three different temperatures, and Figure 4-4-8b shows the Arrhenius plot of the rate constants, from which an activation energy of 71.1 kJ/mol or 17.0 kcal/mol was determined.

Possible Modifications of Procedure

Some modifications to this experiment have been noted by those in other teaching laboratories. In particular, Professor Joseph J. BelBruno, Department of Chemistry, Dartmouth College (personal communication) recommends the following modifications:

1. The experiment is performed cooperatively, and each laboratory group is responsible for one of the four sets of conditions from the following table:

Glycidol Concentration (%)	15μL of (HClO₄), M	T, °C
30	6	30
30	6	40
30	6	50
30	3	50

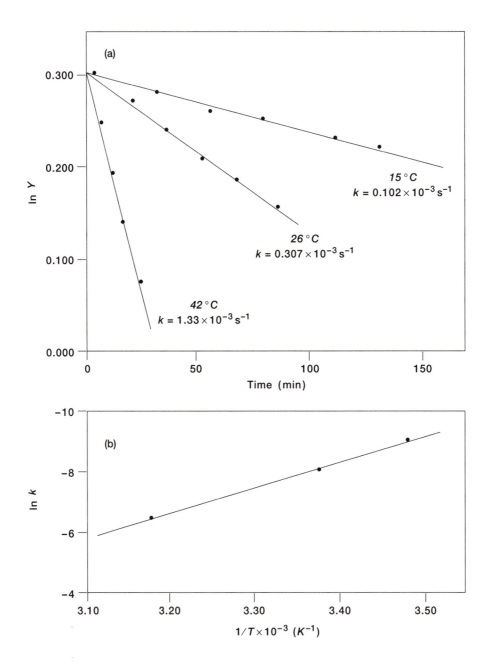

FIGURE 4-4-8. Dependence of pseudo–first-order rate constant (k) on temperature. Plot **a** shows displacement vs. time for the reaction at three different temperatures; plot **b** shows the Arrhenius plot of the rate constants.

The kinetic results of each group including the first-order rate constant are shared by all the groups, and statistical calculations are performed.

2. A thermostating finger and thermocouple are lowered into the reaction cell to monitor and hold stable the temperature of the reaction mixture. The feature circumvents the previously described use of ice in a cooling bath. Because of the smaller volumes of perchloric acid added (and correspondingly longer reaction times), the temperature change is less in this modification.

Further Study

The laser-based detection system described in this experiment could be applied to the study of many reactions in which the refractive index changes significantly during the course of the reaction. Note that the linearity of the refractive index of the reaction mixture versus the extent of reaction must be ascertained for each reaction using a procedure similar to that described in Appendix 4-4-A1. For a three-component system such as the one in this experiment, one method of ensuring this linearity is to maintain a large excess of one of the reactants. Unfortunately, this method simultaneously decreases the change in refractive index during the course of the reaction.

Many kinetic studies of epoxide reactions have been reported (4–7) and some of these systems might be appropriate for study with the laser-based refractive index detection system described in this experiment. Pritchard and Long (6) report that the rate constants of the acid-catalyzed ring openings of several epoxides are larger in D_2O (heavy water) than in H_2O. They did not study the ring opening of glycidol. In addition, Brönsted et al. (4) report that the conversion of glycidol to glycerol increases linearly with total electrolyte concentration (see Appendix 4-4-A3). The effects of adding D_2O and electrolytes to the reaction mixture could easily be studied with the apparatus described in this experiment.

REFERENCES

1. Spencer, B., Zare, R.N. *J. Chem. Ed.* 65 (1988) 835.
2. Streitwieser, A. Jr., Heathcock, C.H. *Introduction to Organic Chemistry,* 3rd ed. Macmillan; New York: 1985; 219–220.
3. Noll, E.D. *Phys. Teach.* 11 (1973) 307.
4. Brönsted, J.N., Kilpatrick, M., Kilpatrick, M. *JACS* 51 (1929) 428.
5. Pritchard, J.G., Long, F.A. *JACS* 78 (1956) 2667.
6. Pritchard, J.G., Long, F.A. *JACS* 78 (1956) 6008.
7. Pritchard, J.G., Long, F.A. *JACS* 80 (1958) 4162.

Refractive Index of the Reaction Mixture as a Function of the Extent of Reaction

If a, b, and c are defined as the number of moles of glycidol, water, and glycerol, respectively, at any given time, then

$$X_2 = \frac{b}{a+b+c} \tag{4-4-A1-1}$$

and

$$X_3 = \frac{c}{a+b+c} \tag{4-4-A1-2}$$

We can also define a_o, b_o, c_o as the initial number of moles of each species and a parameter z to represent the extent of the reaction according to

$$z = \frac{c}{a_o} = \frac{a-a_o}{a_o} = \frac{b_o-b}{a_o} \tag{4-4-A1-3}$$

Note that z ranges from zero, before any of the glycidol has reacted, to one, if the reaction goes to completion. Substituting these expressions into equations 4-4-A1-1 and 4-4-A1-2 gives the mole fractions of water and glycerol as a function of the extent of the reaction:

$$X_2 = \frac{b_o - a_o z}{a_o(1+z)+b_o} \tag{4-4-A1-4}$$

and

$$X_3 = \frac{a_o z}{a_o(1+z)+b_o} \tag{4-4-A1-5}$$

These two expressions can then be substituted into equation 4-4-5 to obtain an analytical expression for the index of refraction of the reaction mixture in terms of the parameter z, the initial number of moles of each reagent, and their indices of refraction:

$$n = n_1 + (n_2 - n_1)\left[\frac{b_o - a_o z}{a_o(1+z)+b_o)}\right] + (n_3 - n_1)\left[\frac{a_o z}{a_o(1+z)+b_o)}\right] \tag{4-4-A1-6}$$

Figure 4-4-A1-1 shows a graph of this function evaluated using the reaction conditions specified in the procedure. Note that the function is very nearly linear over the entire course of the reaction.

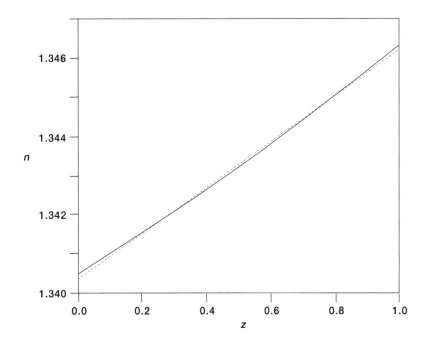

FIGURE 4-4-A1-1. Refractive index of the reaction mixture vs. extent of reaction. The solid line is calculated from equation 4-4-A1-6. The dashed line is a straight line for comparison.

Beam Displacement as a Function of Refractive Index

Noll (1) demonstrates that when the angle θ (Figure 4-4-A2-1) reaches a maximum value, the index of refraction of the liquid is related to the angle of refraction (r) by

$$r = \sin^{-1}\left(\frac{4 - n^2}{3n^2}\right)^{1/2} \qquad (4\text{-}4\text{-}A2\text{-}1)$$

From triangle DCE in Figure 4-4-A2-2, it can be seen that

$$\frac{\theta}{2} + (180° - r) + (i - r) = 180° \qquad (4\text{-}4\text{-}A2\text{-}2)$$

which can be rearranged to

$$\theta = 4r - 2i \qquad (4\text{-}4\text{-}A2\text{-}3)$$

and from Figure 4-4-2

$$\tan\theta = \frac{Y}{X} \qquad (4\text{-}4\text{-}A2\text{-}4)$$

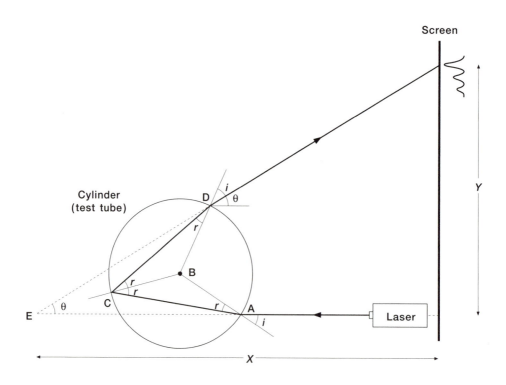

FIGURE 4-4-A2-1. Path of the laser beam through the reaction cell. r is the angle of refraction, and i is the angle of incidence as defined by Snell's law.

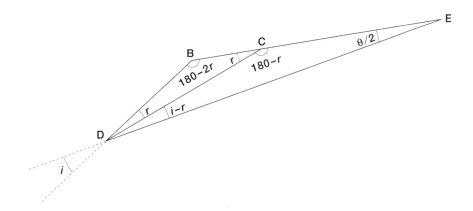

FIGURE 4-4-A2-2. Elucidation of geometrical relationships in circled portion of Figure 4-4-A2-1.

The distance X does not change significantly during the course of the reaction, hence

$$Y \propto \tan \theta \qquad \text{(4-4-A2-5)}$$

Thus, the displacement measured during the course of the experiment is related to the index of refraction of the mixture by

$$Y \propto \tan \theta \left[4 \sin^{-1} \left(\frac{4-n^2}{3n^2} \right)^{1/2} - 2i \right] \qquad \text{(4-4-A2-6)}$$

This complex relationship appears forbidding, but it turns out that for reasonable values of n ($1.33 \leq n \leq 1.40$) and i ($60° \leq i \leq 70°$), Y varies linearly with n (Figure 4-4-A2-3).

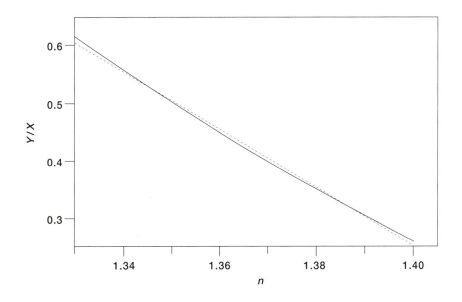

FIGURE 4-4-A2-3. Beam displacement as a function of refractive index. The solid line is calculated from equation 4-4-A2-5. The dashed line is a straight line for comparison.

Dependence of the Rate Constants on Acid Concentration

In strongly acidic media, such as greater than 0.1-M $HClO_4$, the concentration of acid is not an accurate indicator of the hydrogen ion activity (2). A variety of empirically determined "acidity functions" have been developed to relate the concentrations of strong acids in a variety of media to the hydrogen ion activities. The acidity function useful for interpreting the results of this experiment is called the *Hammett acidity function* (H_o). This acidity function determines the hydrogen ion activities of acidic solutions by using indicators whose absorption spectra vary in a regular manner with the hydrogen ion activity (2). The value of H_o that results from this analysis should be directly proportional to the hydrogen ion activity. Charts of $HClO_4$ concentration vs. H_o are available in the literature (3,4), but unfortunately they are not helpful for the range of acid concentrations employed in this experiment.

Several experiments have investigated the dependence of the reaction rate of acid-catalyzed epoxide ring openings on acid concentration (5). Brönsted et al. (6) investigated the conversion of glycidol to glycerol using acid concentrations less than 0.02 M, well below the range of acidities employed in this experiment, and found that the rate constants increased linearly with acid concentration. Pritchard and Long (7) investigated the same reaction at much higher acid concentrations (greater than 0.6 M, much greater than in this experiment), and found the rate constants to be linear with H_o. The range of acidities employed in this experiment falls in between these two experiments, and Figure 4-4-7 demonstrates that at lower acid concentrations, the relationship between the rate constants and acid concentration is approximately linear but that deviations from this linear relationship increase with increasing acid concentration. Presumably the rate constants would be directly proportional to H_o if values of H_o were available for the acid concentrations employed. However, the hydrogen ion activity also depends strongly on total electrolyte concentration (because ionic strength affects hydrogen ion activity), and linearity between H_o and the reaction constants likely will be observed only if care is taken not to contaminate the reaction mixture with any ions other than from the perchloric acid.

REFERENCES

1. Noll, E.D. *Phys. Teach.* 11 (1973) 307.
2. Jones, R.A.Y. *Physical and Mechanistic Organic Chemistry.* Cambridge University Press; New York: 1979; 73–80.
3. Hammett, L.P., Deyrup, A.J. *JACS* 54 (1932) 2721.
4. Rochester, C.H. *Acidity Functions.* Academic; New York: 1970; 43.
5. Rochester, C.H. *Acidity Functions.* Academic; New York: 1970; 138–140.
6. Brönsted, J.N., Kilpatrick, M., Kilpatrick, M. *JACS* 51 (1929) 428.
7. Pritchard, J.G., Long, F.A. *JACS* 78 (1956) 2667.

CHAPTER 5

THE ELECTRONIC STRUCTURE OF MATTER

This chapter contains five experiments that use lasers to study the electronic structure of molecules. Experiments 5-1 through 5-4 involve the study of laser-induced fluorescence from chlorophyll solutions[1] that can be prepared easily as described in experiment 5-1. Experiment 5-5 describes the use of a laser pendulum apparatus to study the magnetic properties of various inorganic compounds and provides insight into the electronic structure of these compounds. The phenomena studied in these experiments are best understood in terms of quantum mechanics. For this reason, this introduction describes simply and concisely some of the key results of quantum mechanics as they pertain to molecules.

Although performing the experiments in no way requires an understanding of quantum mechanics, the concepts presented here are helpful for gaining insight into the physical processes underlying the experiments. In particular, the information in "Molecular Electronic Structure" provides a theoretical basis for the phenomena studied in experiments 5-1 through 5-4. This introduction concludes with a brief discussion of how lasers work, particularly the He–Ne laser, which is used for most of the experiments in this book.

A NEW UNDERSTANDING OF LIGHT AND MATTER

The twentieth century has witnessed a revolution in the scientific understanding of light and matter and how they interact. This scientific revolution led to the discovery of the photon and to the development of the atomic model. Just as important, the laws of physics established before the twentieth century were shown to be unable to account for the existence of an atom or a photon of light.

---

1. We thank Professor Allan L. Smith, Department of Chemistry, Drexel University, Philadelphia, PA, for suggesting to us that He–Ne excitation of chlorophyll solutions might make a good system for studying laser-induced fluorescence.

New concepts were needed to explain physical interactions on the scale of atoms and molecules. This new science became known as *quantum mechanics,* and it still provides the foundation of our current understanding of atoms and molecules.

Before this century, one key assumption made by physicists was that matter and energy were clearly distinct entities. Matter consisted of particles, and radiant energy existed in the form of waves. In the early twentieth century, a series of experiments led scientists to doubt the long-accepted strict distinction between matter and energy. These experiments indicated that light could exhibit particlelike properties and matter could exhibit wavelike properties.

The most sophisticated treatment of the nature of light before the twentieth century was provided by James Clerk Maxwell in the 1870s. Maxwell's equations for electromagnetic radiation state that all forms of radiation including visible light move through space as oscillating electric and magnetic fields, and that these fields arise from acceleration or deceleration of charges. Further, all electromagnetic radiation travels through a vacuum at the same speed, namely 3.0×10^8 m/s (customarily represented by the letter c), and the product of wavelength λ (m) and frequency ν ($s^{-1} = $ Hz) gives the speed of light, that is,

$$c = \lambda \, \nu \qquad\qquad (5\text{-}1)$$

Thus, the Maxwell equations treat light as a form of energy that travels through space as electromagnetic waves.

The first experiment to cast doubt on this model of light was Max Planck's 1900 study of blackbody radiation, the radiant energy emitted from a solid when it is heated. Planck found that he could accurately model his data only if he assumed that the radiation was emitted in whole-number multiples of certain well-defined amounts of energy. To explain this mathematical result physically, Planck created a model in which electromagnetic radiation could be absorbed or emitted only in discrete quantities, the smallest of which was named a *quantum.* From his data, Planck found that the energy of a quantum ΔE (in J) is directly proportional to the frequency ν of the radiation, which he expressed mathematically as

$$\Delta E = h \, \nu \qquad\qquad (5\text{-}2)$$

The proportionality constant h, now known as Planck's constant, was found experimentally to have a value of 6.63×10^{-34} J·s.

Albert Einstein's later studies of the photoelectric effect helped confirm Planck's finding that light carried discrete quantities of energy. In fact, Einstein's experiments indicated that light consisted of a stream of particles, which Einstein named *photons.* This view of light was somewhat unnerving at the time because light clearly did have wavelike properties, such as diffraction effects (see Chapter 3).

At the same time that the understanding of light was changing radically, equally important changes were being made in the understanding of the structure of matter. A series of experiments by Rutherford and others demonstrated that matter was composed of atoms, which consisted of an extremely small, massive nucleus composed of protons and neutrons and surrounded by electrons. The distribution of the electrons around the nucleus was not well understood for several years, however.

Neils Bohr presented the first successful model of the electronic structure of atoms by proposing that electrons in hydrogen atoms travel around the nucleus in circular orbits, much like the moon orbits around the Earth. Additionally, Bohr assumed that these circular orbits could have only certain radii, implying that the hydrogen atom could exist only in certain energy states (often referred to as *quantized energy states*). Bohr's model was a major breakthrough because it accurately predicted the electronic structure of hydrogen. Bohr could not extend his model to atoms other than hydrogen, however, and many experiments demonstrated that Bohr's planetary model of electronic motion is incorrect.

A final important discovery of the early twentieth century was that electrons have a property known as *spin,* which gives rise to a small *magnetic moment,* and that electrons may have either of exactly two types of spin, which are called *spin up* and *spin down.* For more on electron spin, please see experiment 5-5.

AN INTRODUCTION TO THE THEORY OF QUANTUM MECHANICS

In the 1920s, Erwin Schrödinger (1887–1961) and several other scientists developed abstract mathematical theories that have become the foundation of modern quantum mechanics. Unlike Bohr's theory, which could accurately model only the hydrogen atom, Schrödinger's equation appears to model accurately the electronic structure of all atoms and molecules. The key result of quantum mechanics for chemistry is that the internal energy of atoms and molecules is quantized. Similar to Bohr's model of the hydrogen atom, quantum mechanics predicts that the electrons in all atoms and molecules can exist only in certain well-defined energy states. An atom or molecule can undergo transitions between quantized energy states by the absorption or emission of one photon of light with an energy equal to the difference in energy between the two quantized energy levels. The experimental study of the absorption and emission of radiation by atoms and molecules is known as *spectroscopy.*

Unlike the Bohr model for the hydrogen atom, quantum mechanics does not predict circular orbits for electrons. In fact, quantum mechanics does not predict any regular patterns of motion *(trajectories)* for electrons; instead, it predicts only the probabilities of finding electrons in a certain location in space at a certain time. Abandoned was the idea of a classical trajectory (or orbit) in which the position and momentum (mass times velocity) of the electron is known at each instant. Instead, a fuzzy probabilistic picture for the electron emerged in which orbits were replaced by *orbitals,* a probability density of where the electron would be found if its position were measured.

Quantum mechanics can be employed to determine the quantized electronic energy levels of any atom or molecule if the potential energy of the system is known. The orbitals and energy levels of multielectron atoms are well discussed in general chemistry textbooks and will not be discussed here. Instead, we turn to the electronic structure of molecules, which provides a theoretical basis for experiments 5-1 through 5-4.

MOLECULAR ELECTRONIC STRUCTURE

Molecules can store internal energy in several different forms, all of which are quantized.[2] The three most important forms of energy in molecules are *electronic energy, vibrational energy,* and *rotational energy* (Figure 5-1). Electronic energy includes the attraction of each electron to each nucleus and the electron–electron repulsions. Electronic energy levels for molecules generally are separated by energies that correspond to visible or ultraviolet light. Vibrational energy arises from the interactions between the nuclei. The bonds in molecules are not rigid but rather behave like springs that vibrate with a frequency proportional to the energy in the bond. Vibrational energy is also quantized, and transi-

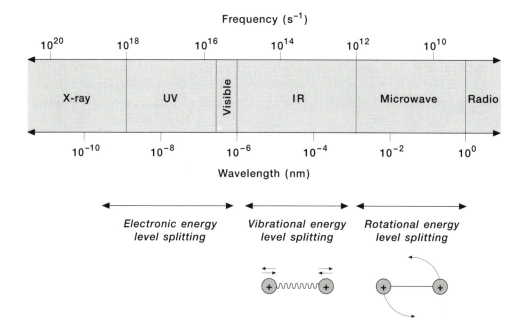

FIGURE 5-1. Electromagnetic spectrum showing location of typical electronic, vibrational, and rotational energy levels.

2. These types of energy are all interlinked because they all depend on the positions of the electrons and nuclei. An excellent approximation (called the Born-Oppenheimer approximation) is to treat each energy form separately and to solve for the quantized energy levels corresponding to each type of motion.

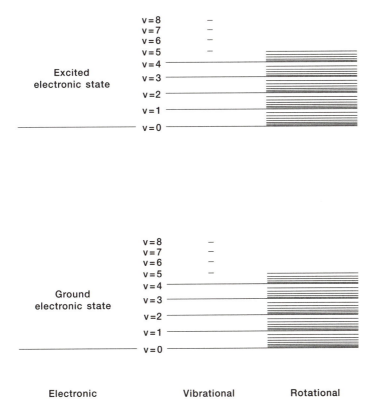

Excited
electronic state

Ground
electronic state

Electronic   Vibrational   Rotational

FIGURE 5-2. Schematic energy-level diagram for a diatomic molecule.

tions between vibrational energy levels are usually observed in the infrared region of the electromagnetic spectrum, which is lower in energy than the visible and ultraviolet regions. Rotational energy arises from the rotation of the molecule about its center of mass. The energy-level spacings of the quantized rotational levels are equivalent to the energies of microwave radiation, even lower in energy than infrared radiation. A typical energy-level diagram for a diatomic molecule[3] is shown in Figure 5-2. Note that each electronic state is subdivided into vibrational energy levels, which in turn are subdivided into even smaller rotational energy levels.

Molecules may have either an odd or even number of electrons. Molecules that have an odd number of electrons are referred to as radicals and tend to be chemically unstable, that is, highly reactive. Most chemically stable molecules have an even number of electrons. In these molecules, electrons are almost always paired, such that for every electron with spin up, there is one electron with spin down. Such molecules are said to be in a *singlet state*. Molecules may also

3. Diatomic molecules are simple cases because they require only one quantum number each to describe their vibrational and rotational states. Larger molecules require more quantum numbers to describe their vibrational and rotational states, and the energy-level diagrams are correspondingly more complex.

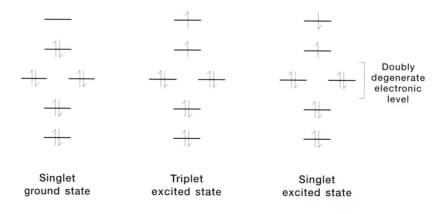

FIGURE 5-3. Electron configurations corresponding to singlet and triplet states.

occupy *triplet states,* in which there are two more electrons with one type of spin than there are with the other type (Figure 5-3). In all cases, the Pauli exclusion principle limits the number of electrons that may occupy any electronic energy level. Many electronic energy levels can hold only two electrons—one with spin up and one with spin down. Other electronic states, known as *doubly degenerate states,* can hold four electrons, two with spin up and two with spin down. Molecules that possess high symmetry can have even higher degeneracies, such as triply degenerate levels that hold six electrons, and so forth.

Electrons often occupy the lowest available electronic energy levels, and when they do, the molecule is said to be in the *ground electronic state.* A molecule in the ground electronic state may or may not be in the ground vibrational and rotational states (in Figure 5-2, v could be greater than 0). At thermal equilibrium, the probability of finding a molecule in a particular vibrational and rotational state depends on the temperature. At higher temperatures, a molecule is more likely to occupy higher energy levels.

Because many different energy-level transitions are possible for molecules, the spectroscopy of molecules is a rich topic. Molecules can change rotational states while they remain in the same electronic and vibrational states, change both vibrational and rotational states, or change all three states. Correspondingly, absorption and emission from molecules occur in many regions of the electromagnetic spectrum. Several experiments in this chapter focus on the phenomenon of *fluorescence,* in which an electron changes from a higher electronic state to a lower one without changing its spin. In general, the rotational and vibrational states also change during such a transition, and the light emitted during fluorescence normally spans a broad range of frequencies.

Typically, for fluorescence to be visible, at least one electron in a molecule must be excited into a higher electronic state, from which it decays to the lower state after some period of time, during which a photon is emitted. A fluorescence experiment in the liquid or solid state can often be thought of as a four-step process, as depicted in Figure 5-4. In this example, the molecule begins in the

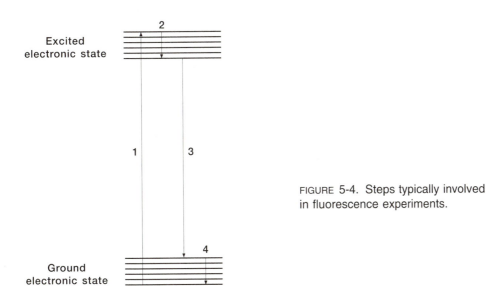

Excited
electronic state

Ground
electronic state

FIGURE 5-4. Steps typically involved
in fluorescence experiments.

ground electronic state. Step 1 corresponds to the absorption of a photon of light, during which one electron enters a higher electronic energy level, and the rotational and vibrational states of the molecule may change as well. In step 2, the molecule remains in the excited electronic state but decays from a higher rotational and vibrational state to a lower one through a nonradiative process. In this step, the molecule does not emit a photon of light but instead transfers energy to surrounding molecules in the form of heat. Step 3 is fluorescence, during which the molecule may or may not return to the original rotational and vibrational state of the molecule. A final step may also occur in which the molecule in the ground electronic state decays from higher rotational, vibrational, or both states to lower ones by a nonradiative process. This diagram shows that the energy changes during absorption and fluorescence are not necessarily equal and that the frequency of the photon emitted during fluorescence is often less than the frequency of the photon absorbed, that is, fluorescence is typically "red shifted" from absorption. Experiment 5-3 provides an example of how to take advantage of this property of fluorescence.

Fluorescence is only one of several processes an excited-state molecule can undergo. Figure 5-5 shows that transitions between singlet and triplet states are also possible. A molecule in an excited singlet state may move to a triplet state by a process known as *intersystem crossing,* in which the spin of one of the electrons flips.[4] In addition, an excited triplet state may decay to a lower energy singlet state (often the ground state) by emitting a photon. This phenomenon is known as *phosphorescence,* and it is responsible for the light emitted from many glow-in-the-dark substances. Molecules in singlet and triplet states also may return to the ground state by nonradiative decay (please see experiments 5-2 and 5-3 for more discussion of this type of nonradiative process).

_____

4. Quantum mechanical calculations demonstrate that the lowest-lying triplet excited state is usually lower in energy than the corresponding lowest-lying excited singlet state, as depicted in Figure 5-5.

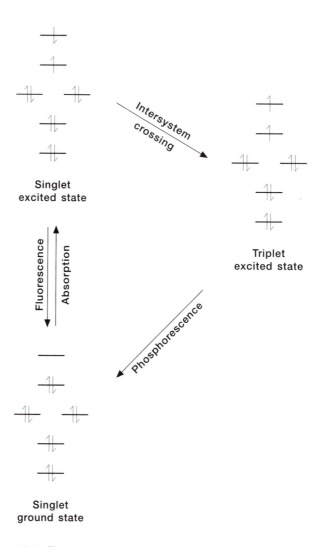

FIGURE 5-5. Transitions among excited states and the ground state.

PRINCIPLES OF LASER OPERATION

The word LASER is an acronym for the rather forbidding phrase **L**ight **A**mplification by **S**timulated **E**mission of **R**adiation. The output of a laser is a beam of light that is bright, single in color, directional, and in phase. The terms *intense, monochromatic, collimated,* and *coherent* are synonymous. From its origins in the late 1950s, the laser has evolved rapidly into a great diversity of available types so that the output of current lasers spans the electromagnetic spectrum from the microwave (maser) to the X-ray.

According to quantum mechanics, atoms and molecules can exist in only discrete, quantized energy levels. Consequently, these building blocks of matter can change their energy levels by absorbing or emitting radiation that has fixed amounts of energy; that is, the radiative transitions in atoms and molecules occur at distinct, sharp frequencies or wavelengths and give rise to line spectra. The laser is a "quantum device" in that its characteristic monochromatic light originates from photons of a single frequency emitted as a result of some radiative transition between energy levels of the device. In fact, most lasers operate between several quantum states to produce a range of different laser frequencies that cover different regions of the electromagnetic spectrum. By proper choice of the optical elements that are an integral part of the laser device, typically one frequency can be made to dominate the others. Stated differently, the laser can be tuned and controlled to produce a beam of monochromatic light that has an extremely narrow frequency width. Sometimes, this width may approach a few hertz (Hz), far less than the frequencies of 10^{15} Hz that characterize light in the visible portion of the electromagnetic spectrum. For example, the helium–neon laser, a gas-filled device that produces a red beam of light when an electrical discharge excites the electrons of the helium and neon atoms inside the laser to higher energy levels, produces light at 632.8 nm in wavelength (4.474×10^{14} Hz in frequency). With the proper choice of optical elements, a helium–neon laser can produce green light at 543.5 nm or several frequencies of infrared light that cannot be sensed by the eye.

How, then, does the laser differ from an atomic lamp or gas discharge tube that also produces light? Many people are familiar with yellow sodium streetlights or mercury-vapor-filled fluorescent lights. In these discharge devices, light is initially produced through spontaneous emission, in which excited species make a radiative transition from higher to lower quantized energy levels on their own. The excitation process is depicted in "before" and "after" diagrams in Figure 5-6. The number 2 denotes the higher level and 1 the lower level; level 0 represents the ground state occupied almost exclusively under normal (thermal)

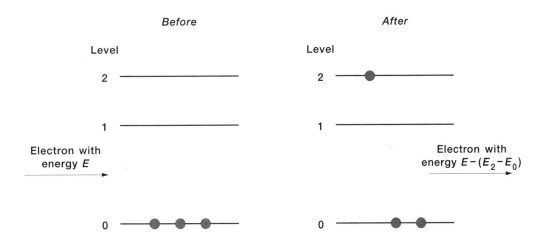

FIGURE 5-6. The two steps "before" and "after" of the excitation of a three-level system by electron impact from the lowest level 0 to the highest level 2. The closed circles symbolize the number of atoms or molecules that occupy a quantum level.

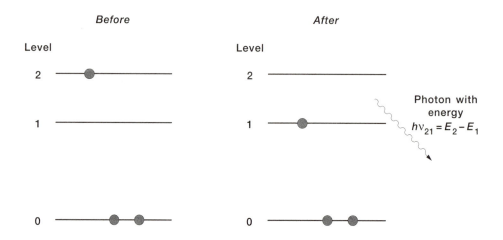

FIGURE 5-7. The two steps "before" and "after" of spontaneous emission of a three-level system from level 2 to level 1. Note the generation of a photon with energy $E_2 - E_1 = h\nu_{12}$. The closed circles symbolize the number of atoms or molecules that occupy a quantum level.

conditions. Figure 5-7 depicts the "before" and "after" stages of spontaneous emission. The photons produced in the $2 \rightarrow 1$ transition occur spontaneously and have a frequency $\nu_{21} = (E_2 - E_1)/h$, where h is Planck's constant (see equation 5-2).

In a sodium street lamp, more sodium atoms are in the lower state than the upper state. Consequently, photons of yellow light that are spontaneously emitted either escape the sodium vapor in the lamp or become absorbed by some of the unexcited sodium atoms. Suppose, however, that the discharge process is able to create a *population inversion,* in which more atoms or molecules are in level 2 than in level 1. Then, when a photon of the resonant frequency ν_{21} passes by excited atoms or molecules in level 2, it induces them to emit radiation in an inverse process to absorption called *stimulated emission* (see Figure 5-8). In stimulated emission, the emitted photons have the same energy and are in phase with the stimulating photons. Consequently, a traveling wave of electromagnetic energy grows. Because of the population inversion, the system shows gain and acts to amplify the radiant energy of photons with frequency $\nu_{21.}$

A key element in laser operation is the creation of a population inversion. Such a phenomenon cannot be achieved by any amount of heating in an equilibrium manner. At equilibrium, a system at any temperature always obeys Boltzmann statistics for the population of its energy levels. Thus, the population in a level (divided by the degeneracy of that level, i.e., the number of different ways of occupying that level with the same energy) is always larger the lower the energy level is. At thermal equilibrium, no population inversions can occur. Consequently, laser action requires some nonequilibrium process, such as a flash

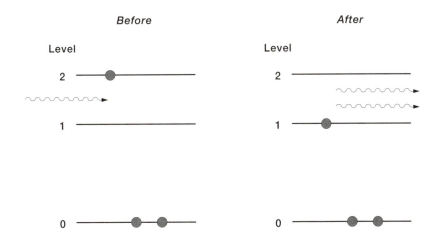

FIGURE 5-8. The two steps "before" and "after" of a three-level system undergoing stimulated emission from level 2 to level 1 by a photon of energy $E_2 - E_1 = h\nu_{12}$. The process produces more photons with the same energy and same phase in wave motion. The closed circles symbolize the number of atoms or molecules that occupy a quantum level.

of light, an electrical discharge, or a chemical reaction (explosion), for population inversion to be produced. If the population inversion is transient, the lasing duration must be finite, and the laser provides a pulsed output. If the population inversion can be maintained, the laser action can occur without time limit, and the output is a continuous wave or, as commonly abbreviated, cw.

Laser action can occur by a single pass of stimulating radiation through an active laser medium, which is characterized by having a population inversion. The result is a *superradiant laser.* For more efficient use of the energy stored in the active medium, however, two highly reflective mirrors are placed on opposite sides of the medium so that the stimulated radiation repeatedly bounces back and forth through the medium between the mirrors, thus increasing the effective pathlength of the active (gain) medium. The use of the optical cavity causes the laser output to have closely spaced optical modes because only an integral number of wavelengths fit inside the length of the optical cavity. For example, the spacing between modes in a helium–neon laser is typically in the megahertz range (10^6 Hz). These modes are so close together that they cannot be distinguished as different colors by the eye. Under special circumstances, only one mode may be selected to lase to produce a so-called *single-mode laser.* In the next section, we describe the helium–neon laser, a commonly used and well-known laser whose output in the red may be used to carry out most of the experiments described in this book.

THE HELIUM–NEON LASER

In 1961, Javan, Bennett, and Herriott of Bell Telephone Laboratories reported demonstrating the first continuous action laser, which was achieved by maintaining an electrical discharge in a gas mixture of helium and neon atoms (1). The lasing cavity of a typical helium–neon laser consists of a sealed tube filled with helium gas (90%) and neon gas (10%) at a total pressure of approximately 1 torr (equivalent to 1/760 of atmospheric pressure). At this low pressure, a high-voltage power supply easily maintains an electrical discharge inside the sealed cavity, thereby producing energetic electrons that travel through the gas mixture. The typical current is 10 milliamperes at 1,000 V, and the typical power is 10 W, which produces a 1-mW output beam for an overall efficiency of only 0.01% for converting electrical to coherent radiant energy.

The atomic energy levels of the rare gases are a complicated topic. The ground state of a rare gas such as helium or neon has a closed-shell configuration of electrons, $1s^2$ for He and $1s^22s^22p^6$ for Ne, each of which gives rise to only one level, denoted 1^1S. Excited electron configurations of the rare gases have more than one open shell of electrons and give rise to numerous atomic energy levels. To confuse matters, the standard spectroscopic notation (2) is based more on historical than logical reasons, and the labels are best considered names for the levels that occur at certain energies.

Figure 5-9 presents the major energy levels of He and Ne that are relevant to understanding how the He–Ne laser works. The electrons in the discharge excite helium atoms from the $1s^2$ ground-state electron configuration to the $1s2s$ excited-state electron configuration, which has two long-lived (metastable) atomic energy levels, denoted 2^1S and 2^3S. The 2^1S level lies very close in energy to atomic energy levels that belong to the $1s^22s^22p^55s$ electron configuration of neon. As a consequence, the excited 2^1S helium atoms efficiently transfer by collision their energy to neon atoms to populate the upper level (denoted $3s_2$) of the 632.8-nm laser transition (corresponding to $3s_2 \rightarrow 2p_4$). In truth, this collisional transfer process is much more efficient for exciting this level of neon than is direct electron-impact excitation of neon. Thus, the He–Ne laser works by photosensitization of Ne atoms by excited He atoms; Chapter 6 introduces a number of other examples of photosensitized energy transfer.

Because only the ground-state electron configuration $1s^22s^22p^6$ of neon is populated at room temperature, a population inversion is created between this upper level of the $1s^22s^22p^55s$ configuration of Ne and the lower-lying levels of the $1s^22s^22p^53p$ electron configuration of Ne. Laser action at 632.8 nm can then occur if the excited gas mixture is placed between suitable mirrors, which are commonly sealed directly to the ends of the discharge tube. As shown in Figure 5-9, 3.391-μm laser action (in the infrared) competes with the 632.8-nm laser action (in the visible) because both of these laser transitions deplete the same upper level. Increasing the intensity of the 632.8-nm laser line necessitates "spoiling the gain" of the 3.391-μm transition by using, for example, an infrared-absorbing mirror.

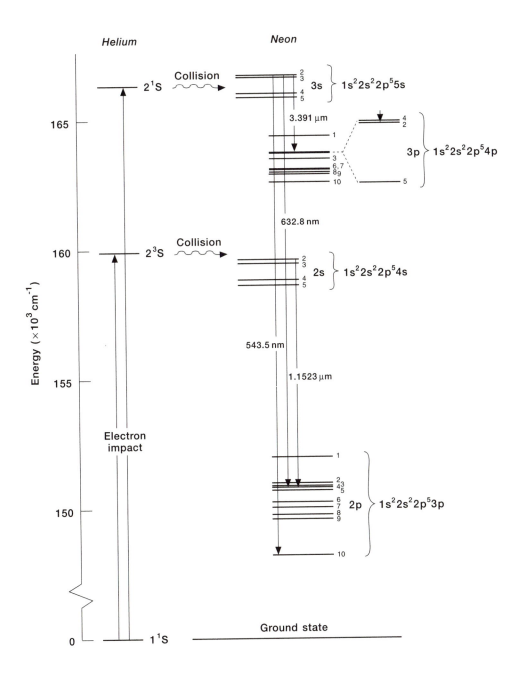

FIGURE 5-9. An energy-level diagram of helium and neon, showing the excitation mechanism and neon transitions responsible for laser action in a helium–neon laser.

The levels of the $1s^2 2s^2 2p^5 3p$ configuration of neon rapidly decay radiatively to levels of the $1s^2 2s^2 2p^5 3s$ configuration (not shown in Figure 5-9), primarily by transitions in the red portion of the spectrum. This rapid decay gives neon signs their characteristic color. This efficient emptying of the population of the lower level of the 632.8-nm laser transition is very desirable because it ensures that the population inversion between the upper and lower levels of this transition is continuously maintained as long as the electric discharge occurs in the gas mixture. Neon atoms in the long-lived levels of the $1s^2 2s^2 2p^5 3s$ configuration are returned to the ground-state configuration by deactivating collisions, which include collisions with the walls of the discharge tube.

The helium–neon laser transition at 632.8 nm has rather low gain, typically 2–10% per meter. Consequently, one end mirror is nearly 100% reflective, whereas the other is 99% reflective, a level that permits only 1% of the intracavity radiation to escape per round trip inside the cavity. To support laser action, the optical cavity must be carefully aligned so that the light beam inside the cavity does not escape ("walk off") the end mirrors after repeated reflections. As a He–Ne laser is turned on, it warms up, and the dimensions of its cavity structure change over time. Typical He–Ne lasers operate in a multimode manner, and the intensity of the different modes varies as the cavity structure changes with temperature. Consequently, the He–Ne laser beam varies in intensity with time, particularly during the warm-up stage. This intensity variation can be readily detected and displayed using a power meter that records the light intensity striking its surface. Variations of several percent are typical for inexpensive, unstabilized He–Ne lasers.

A helium–neon laser can show laser action (oscillation) on more than 50 different transitions. The first He–Ne laser built by Javan, Bennett, and Herriott (1) operated at 1.1523 μm, using the $2s_2 \rightarrow 2p_4$ transition in the near infrared range (see Figure 5-9). Shortly thereafter, White and Rigden (3) discovered laser action in the red 632.8-nm transition; this red beam of coherent light is now familiar to us in the form of bar code readers at check-out counters and laser pointers. The He–Ne laser may also be made to operate in the green at 543.5 nm (see Figure 5-9) using the $3s_2 \rightarrow 2p_{10}$ transition.

More information on lasers may be found in references 4–9.

REFERENCES

1. Javan, A., Bennett, W.R. Jr., Herriott, D.R. *Phys. Rev. Lett.* 6 (1961) 106.
2. Moore, C.E. *Atomic Energy Levels,* vol. I. U.S. Government Printing Office; Washington, DC: 1949; 5 (for He); 77 (for Ne). NSRDS-NBS 35.
3. Rigden, J.D., White, A.D. *Proc. IRE* 50 (1962) 2366.
4. O'Shea, D.C., Callen, W.R., Rhodes, W.T. *Introduction to Lasers and their Applications.* Addison-Wesley; Reading, MA: 1978.
5. McComb, G. *Laser Cookbook.* Tab Books; Blue Ridge Summit, PA: 1988.
6. Metrologic Instruments Catalogue, Coles Road at Rte. 42, Blackwood, NJ 08012.
7. Polik, W.F. In Schwenz, R.W., Moore, R.J., eds. *Physical Chemistry: Developing a Dynamic Curriculum.* American Chemical Society; Washington, DC: 1993. Chapter 6: 85–108.
8. Kallard, T. *Exploring Laser Light.* Optosonic Press. Reprinted by American Association of Physics Teachers, 5110 Roanoke Pl., Ste. 101, College Park, MD 20740: 1977.
9. Siegman, A.E. *Lasers.* University Science Books; Mill Valley, CA: 1986.

EXPERIMENT 5-1

EXTRACTION OF CHLOROPHYLL FROM FRESH SPINACH

Experiments 5-2, 5-3, and 5-4 study the fluorescent properties of chlorophyll. In this experiment, chlorophylls a and b are extracted from fresh spinach and laser-induced fluorescence from these compounds is observed.

DEGREE OF DIFFICULTY

Experimental: moderate
Conceptual: easy

MATERIALS

- Erlenmeyer flasks
- beakers
- mortar and pestle
- separatory funnel, 125 mL
- round-bottom flask, 150 mL
- rotovap or steam bath
- column chromatography equipment (buret is sufficient)
- glass wool
- 10 g of alumina powder (aluminum oxide)
- 10 g of anhydrous magnesium sulfate
- 150 mL of heptane (hexane is an acceptable substitute) (Sigma product #H9629. Approx. $18 for 1 liter)
- 5–10 fresh spinach leaves
- 10 g of sand
- He–Ne laser (or other laser with wavelength very near 633 nm)
- cut-off filter, 665 nm or 645 nm (if available) (Available from Corion Corp., 73 Jeffrey Ave., Holliston, MA 01746-2082. Listed as "Schott filter glass," LG-650 or LG-675. Approx. $20 for a 1-in.-diameter filter)

PROCEDURE

Extraction of Chlorophyll from Spinach Leaves

1. Grind 4 or 5 large spinach leaves in a pool of heptane using a mortar and pestle. A pinch of sand may aid the process. Pour off the green organic extract, and filter it using a piece of filter paper and a funnel.

2. To remove impurities, extract the heptane solution with water in a separatory funnel. Repeat the washing procedure, and then pour the green heptane solution into an Erlenmeyer flask with 10 g of anhydrous magnesium sulfate. Let the solution sit for several minutes.

3. Pour off the heptane solution into a round-bottom flask. Concentrate the solution in the flask by heating it in a water bath at a temperature below 50° C (by swirling or using a rotovaporator) until approximately 1 mL of the solution remains. Reduce the temperature to below 40° C, and allow the solution to evaporate to dryness.

4. Dissolve this residue with 2 mL of heptane and set aside.

Preparation of the Alumina Column

5. Place a small amount of glass wool at the bottom of a chromatography column, and fill the column two-thirds full with heptane. Make a one-half-inch plug of sand above the glass wool by slowly pouring in sand and allowing it to settle to the bottom.

6. Slowly pour in alumina powder while tapping on the sides of the column to prevent clumping and bubble formation. Make sure the liquid level never falls below the column packing. Drying of the column leads to cracking, which makes the column perform poorly. After all of the alumina has been added, open the stopcock and allow the heptane to drain slowly to allow the column to settle. Drain the heptane until it just covers the packing.

Purification of the Chlorophyll

7. Pipette the green extract from step 4 onto the top of the column, and drain the column slowly until the alumina is nearly exposed. Add 5 mL of heptane to the column, and allow it to drain as well. Add another 5 mL of heptane and a small amount of sand to the column. Fill the column to the top with heptane. Caution: never let the liquid level fall below the level of the alumina.

8. Drain the solvent through the column and collect the various fractions (the colored bands that pass through the column) in separate flasks. You may wish to weigh several of the flasks in advance to ease preparation of the chlorophyll solutions in step 9. The yellow–green carotenes come off first followed by the two chlorophylls, a and b (green). Experiments 5-2 through 5-4 use the chlorophyll fractions. Consequently, do not contaminate any of the green fractions with the yellow carotenes. Distinguishing between chlorophyll a and b is not necessary.

9. Evaporate the chlorophyll fractions to a green residue without heating them above 45° C. Determine the mass of each residue. Use these residues immediately to prepare stock solutions of chlorophyll (MW ≈ 900) in heptane. The choice of concentration of these solutions is up to the experimenter; we prepared stock solutions of concentration 1.30×10^{-3} M. These solutions should be capped and refrigerated until needed.

10. Irradiate the chlorophyll solutions of varying concentrations with a red laser, and observe the fluorescence at 90° to the incident laser beam. Confirm that the light arises from fluorescence and not scattering (i.e., that it has a longer wavelength than the incident light) by viewing the cuvette through a 665- or 645-nm cut-off filter.

HAZARDS AND PRECAUTIONS

Heptane is flammable and will irritate the skin upon exposure. Exercise caution when heating heptane solutions; use a boiling stone and a sand bath. Never use flames! Wear gloves when handling these substances, and avoid breathing their vapors. The remaining substances are believed to be of low toxicity and present no hazards known at this time. Wear eye protection at all times.

DISPOSAL

All solids can be disposed of as trash. Any heptane solutions should be disposed of as organic waste.

DISCUSSION

The column chromatography system described in the procedure is conceptually identical to the column prepared in experiment 4-1. The laser-based detection system is unnecessary in this experiment because the desired compounds are colored. Please refer to the discussion section of experiment 4-1 for background information on chromatography.

Figure 5-1-1 shows the structures of chlorophylls a and b. These molecules are examples of coordination compounds, which have in common a metal ion coordinated to an organic compound. Coordination compounds include such diverse compounds as hemoglobin and vitamin B_{12} and are essential in many biological processes, such as the transport of oxygen, electron transfer, and catalysis.

Chlorophyll is the component in photosynthesis that traps solar energy so that it can be used to drive the production of carbohydrates from carbon dioxide and water. Chlorophyll a and b absorb strongly in the red region of the spectrum and have peak

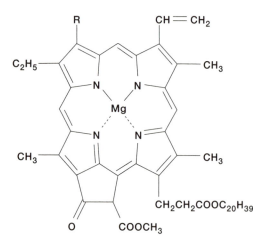

FIGURE 5-1-1. Molecular structure of chlorophyll. When R = CH_3, the compound is chlorophyll a; when R = CHO, the compound is chlorophyll b.

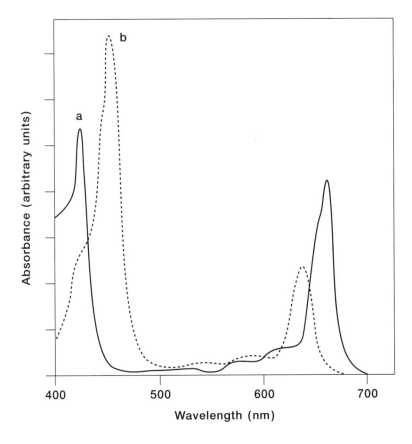

FIGURE 5-1-2. Absorption spectra of chlorophylls a and b.

absorptions near 650 nm (Figure 5-1-2). The absorption spectra for chlorophylls a and b differ somewhat, and these two types of chlorophyll complement each other in absorbing sunlight.

During photosynthesis, once a photon is absorbed by a chlorophyll molecule, the energy from that absorption is quickly transferred to the chloroplasts, where it is used to drive chemical processes. When the chlorophyll is extracted from leaves, however, the absorbed energy can no longer be trapped, and the chlorophyll reemits the energy by fluorescence. The chlorophyll solutions prepared in this experiment strongly absorb the 632.8-nm light from a He–Ne laser, and the fluorescence maximum lies at approximately 685 nm. The fluorescence can be distinguished from scattered laser light by use of a filter. Unlike many synthetic dyes, chlorophyll is nontoxic, plentiful, and cheap. For these reasons, chlorophyll is an ideal substance for studying absorption and fluorescence phenomena in the laboratory.

EXPERIMENT 5-2

FLUORESCENCE VS. CONCENTRATION

This experiment examines the intensity of fluorescent light in a series of solutions of different chlorophyll concentrations to determine the dependence of fluorescence intensity on concentration.

DEGREE OF DIFFICULTY

Experimental: moderate
Conceptual: moderate

MATERIALS

- He–Ne laser (or other laser with wavelength very near 633 nm)
- cuvette and holder
- light-detection system
- test tubes, graduated cylinders, or volumetric flasks
- Schott cut-off filters, RG665 or RG645 (Available from Corion Corp., 73 Jeffrey Ave., Holliston, MA 01746-2082. Listed as "Schott filter glass," LG-650 or LG-675. Approx. $20 for a 1-in.-diameter filter)
- 0.45-μm filter (if available)
- chlorophyll solution (from experiment 5-1)
- heptane (Sigma product #H9629. Approx. $18 for 1 liter)
- beam splitter (if available)

PROCEDURE

1. Assemble the apparatus as described in step 1 of experiment 5-3 and depicted in Figure 5-3-1.

2. Prepare six to eight dilutions of chlorophyll (MW $\approx$ 900) from the stock solution (prepared in experiment 5-1) and heptane. Recommended concentrations are from 1.3×10^{-3} M to 4.0×10^{-5} M. Prepare at least 3 mL of each chlorophyll solution. Filter the solutions with a 0.45-μm filter, if available.

3. Fill the cuvette with the most dilute sample and irradiate it with red laser light. Record the fluorescence intensity.

4. Repeat this procedure using the other dilutions. When changing solutions, empty the cuvette, and rinse it with a small amount of the new solution from a clean pipette before filling the cuvette with the new solution. Be sure not to move the sample holder when changing solutions.

5. Prepare a plot of fluorescence intensity vs. concentration of chlorophyll.

162

HAZARDS AND PRECAUTIONS

Heptane is flammable and will irritate the skin upon exposure. Keep all of the organic solutions away from open flames, wear gloves when handling them, and avoid breathing their vapors. The other substances are believed to be of low toxicity and present no hazards known at this time. Wear eye protection at all times.

DISPOSAL

The chlorophyll solutions should be disposed of as organic waste.

DISCUSSION

Figure 5-2-1 is a plot of fluorescence intensity as a function of chlorophyll concentration. At low concentration, the intensity rises rapidly and nearly linearly with concentration, whereas at higher concentration, the intensity approaches a constant value.

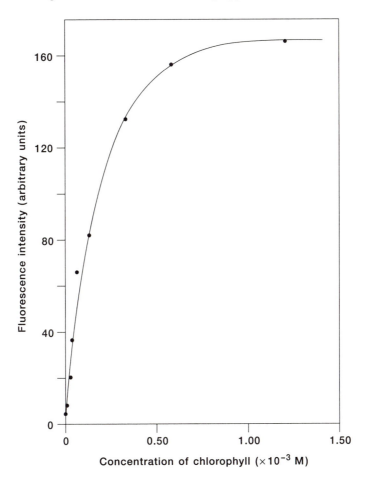

FIGURE 5-2-1. Concentration dependence of chlorophyll fluorescence.

The qualitative form of the concentration dependence of fluorescence for chlorophyll is observed for many different molecules. In fact, the relationship between fluorescence intensity and concentration can be modeled for many systems according to a fairly simple equation:

$$F = \Phi I_o (1 - e^{-\epsilon bc}) \qquad (5\text{-}2\text{-}1)$$

where F is the intensity of fluorescence, I_o is the intensity of incident light, ϵ is the molar absorptivity (a constant for any particular compound at a particular wavelength), b is the pathlength of the cell, and c is the molar concentration. The (dimensionless) constant Φ is the quantum efficiency and has been defined as the number of quanta of radiation fluoresced by a sample divided by the number of quanta absorbed. Thus, the quantum efficiency represents the fraction of molecules that fluoresce from the excited state as opposed to decay by nonradiative processes. This formula accurately predicts that the fluorescence intensity will approach a maximum value ΦI_o as c becomes large.

Formula 5-2-1 also predicts the linear relationship between fluorescence and concentration at low concentration, since it can be shown (using a Taylor series expansion) that

$$e^x \approx 1 + x \qquad (5\text{-}2\text{-}2)$$

for small x (in this case, low concentration). Using this expansion in equation 5-3-1, we find

$$F = \Phi\, I_o\, \epsilon bc \qquad (5\text{-}2\text{-}3)$$

The interested student may wish to use these relationships to model the chlorophyll fluorescence.

FLUORESCENCE QUENCHING

This experiment examines the phenomenon of quenching, which occurs when a substance in solution (the "quencher") interacts with excited-state molecules to decrease the intensity of fluorescent light.

DEGREE OF DIFFICULTY

Experimental: moderate
Conceptual: moderate

MATERIALS

- He–Ne laser (or other laser with wavelength very near 633 nm)
- cuvette and holder
- light-detection system
- test tubes, graduated cylinders, pipettes
- cut-off filter, 665 nm or 645 nm (Available from Corion Corp., 73 Jeffrey Ave., Holliston, MA 01746-2082. Listed as "Schott filter glass," LG-650 or LG-675. Approx. $20 for a 1-in.-diameter filter)
- 0.45-μm filter (if available)
- stock chlorophyll solution (from experiment 5-1)
- heptane
- beam splitter (if available)
- 1 g of sodium iodide (NaI)
- 1 g of p-benzoquinone (Sigma product #B1266. Approx. $20 for 25 g)
- 50 mL of acetone *(optional)*
- 50 mL of propylene glycol (or other high-viscosity solvent) *(optional)*

PROCEDURE

1. Assemble the apparatus as shown in Figure 5-3-1. Optimally, a cuvette holder such

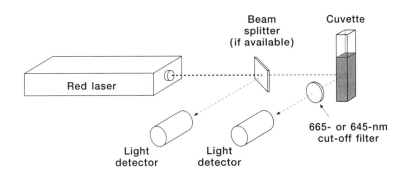

FIGURE 5-3-1. Schematic diagram of experimental setup for detecting fluorescence from chlorophyll solutions.

as the one in Figure 1-3 can be modified to hold the cut-off filter next to the 90° face of the cuvette. Position the photodiode (or other light-measuring device) as close to the cuvette as possible, and make certain that the laser beam passes adjacent to the photodiode. Use the beam splitter to monitor the laser output power, and take measurements only when the power is at some fixed value.

2. Prepare "spiked" solutions of heptane by adding small amounts of NaI or p-benzoquinone. Useful concentrations for these two quenchers are up to 0.5 M for NaI and up to 0.15 M for p-benzoquinone.

3. Evaporate several 2-mL samples of the stock solutions of chlorophyll in heptane (from experiment 5-1) to a green residue without heating beyond 45° C. Dissolve the residues in 2.0 mL of the various spiked solutions. Prepare at least one solution without any quencher as a control. Filter the solutions with a 0.45-μm filter, if available.

4. Measure the fluorescent light intensity for the spiked chlorophyll solutions and the control.

5. (Optional) Repeat the experiment with either acetone or propylene glycol as the solvent to measure the effects of viscosity on quenching.

6. Prepare plots of fluorescence intensity vs. concentration for each quencher and solvent used.

HAZARDS AND PRECAUTIONS

Heptane is flammable and irritates the skin upon exposure. Exercise caution when heating heptane solutions; use a boiling stone and a sand bath. Never use flames! Keep all of the organic solutions away from open flames, and avoid skin contact and breathing their vapors. p-Benzoquinone is a skin and eye irritant and a suspected carcinogen. Wear eye protection at all times.

DISPOSAL

The chlorophyll solutions should be disposed of as organic waste.

DISCUSSION

In addition to fluorescence, several processes can return an excited-state molecule to the ground state without emitting light; these processes are known as *nonradiative processes*. In a typical nonradiative process, the energy of the excited electronic state of a molecule is either transferred to other molecules or redistributed inside the molecule. The mechanisms of these processes are often complex and not well understood. Nonradiative processes compete with fluorescence in the sense that for any given sample of molecules in an excited electronic state, a certain fraction will fluoresce and the rest will return to the ground state through nonradiative processes. The ratio of these two processes is different for every molecule and may be expressed in terms of the fluorescence quantum yield (Φ), which is defined as the number of photons of light fluoresced by a sample divided by the number of photons absorbed. Values of Φ range from nearly 1 for some laser dyes to nearly 0 for molecules that hardly fluoresce.

The quantum yield for a particular molecule depends on the environment of the molecule, that is, the solvent, temperature, pH, and other factors that affect interaction between the excited-state molecule and its surroundings. This experiment examines the effects of two different quenchers that promote nonradiative processes in the chlorophyll solutions. For sodium iodide, the iodide ions in solution interact with the excited-state chlorophyll molecules to promote conversion of the singlet state to a triplet state, which then decays to the ground state by other nonradiative processes.[1] Figure 5-3-2 presents a plot of the extent of fluorescence quenching against concentration of iodide ion.

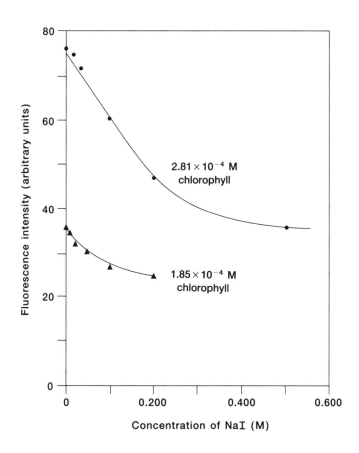

FIGURE 5-3-2. Quenching of chlorophyll fluorescence by sodium iodide.

1. The mechanism for this interaction is that the iodide ion is paramagnetic and has a permanent magnetic dipole moment, which allows it to interact with the electrons in the chlorophyll molecule so as to change the spins of the outermost electrons from spin antiparallel (singlet state) to spin parallel (triplet state). More information on electron spin and magnetic moments is presented in experiment 5-5.

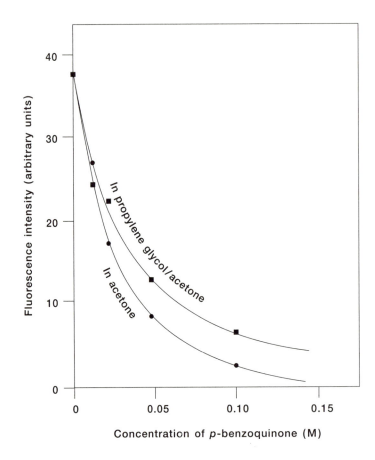

FIGURE 5-3-3. Quenching of chlorophyll fluorescence by *p*-benzoquinone in two solvent systems.

Fluorescence also depends to some extent on the viscosity of the medium. As can be seen in Figure 5-3-3, fluorescence quenching caused by *p*-benzoquinone is greater in the low-viscosity acetone solution than in the higher viscosity propylene glycol solution. In the high-viscosity solution, the chlorophyll and quenching molecules move more slowly, and the probability is lower that an excited-state chlorophyll molecule will come in close contact with a quencher molecule before it fluoresces. Thus, quenching is a less-efficient process in viscous media.

Quenching plays a very important role in photosynthesis. For photosynthesis to be efficient, the potential energy gained by the absorption of photons must be prevented from being reradiated as fluorescence. Quenching occurs during photosynthesis when the chlorophyll comes in contact with a chloroplast, and the energy of the excited state leads to charge separation, which is then used to drive chemical reactions.

EXPERIMENT 5-4

FLUORESCENCE DEPOLARIZATION

The dependence of fluorescence depolarization on temperature is investigated, providing insight into the dynamics of chlorophyll molecules in solution.

DEGREE OF DIFFICULTY

Experimental: moderate
Conceptual: moderate

MATERIALS

- He–Ne laser (or other laser with wavelength very near 633 nm)
- light-detection system
- cuvette holder
- beam splitter (if available)
- water bath and ice bath
- thermometer
- 2 polarizers
- hot plate
- chlorophyll solution (from experiment 5-1)

PROCEDURE

1. Set up the apparatus as shown in Figure 5-4-1. Although not required, continual monitoring of the temperature of the chlorophyll solutions while taking light intensity measurements is particularly convenient. Place one polarizer in the path of the incident laser light and the other parallel to the first in front of the photodiode at 90° to the incident beam. Using a cuvette holder such as the one in Figure 1-3, which is

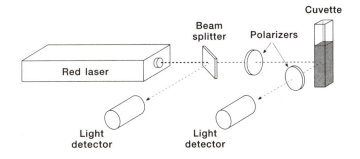

FIGURE 5-4-1. Schematic diagram of experimental setup for detecting depolarization of fluorescence from chlorophyll solutions.

modified to hold the polarizer next to the 90° window of the cuvette, may be helpful. Use the beam splitter to monitor the power output of the laser, and take data only when the power is at some fixed value.

2. Prepare a solution of chlorophyll that gives strong fluorescence upon irradiation by a red laser (see Figure 5-3-1). Measure the light intensity at 90° to the incident beam as a function of the temperature of the chlorophyll solution over a temperature range of 0° to 45° C. One easy way to make these measurements is to heat the solution to 45° C and then measure the fluorescent light intensity as the temperature drops. An ice bath can be used to bring the solution to 0° C. Realize, however, that during the cooling process, the solution is not at thermal equilibrium, and the temperature is not uniform throughout the sample. Use the light meter to monitor the laser output and ensure that all readings are made at the same output.

3. Plot the light intensity at 90° vs. temperature.

HAZARDS AND PRECAUTIONS

Heptane is flammable and irritates the skin upon exposure. Keep all of the organic solutions away from open flames, and avoid skin contact and breathing their vapors. Keep all flammable material away from open flames. Wear eye protection at all times.

DISPOSAL

The chlorophyll solutions should be disposed of as organic waste.

DISCUSSION

In this experiment, ground-state chlorophyll molecules are excited to a higher energy, singlet electronic state by absorption of the light radiation from the laser. This electronic transition occurs by an interaction between the electromagnetic field of the light and the electrons in the molecule. The oscillating electric field of the light causes the electrons in the molecule to oscillate at the same frequency. The additional energy gained by this interaction allows the molecule to enter an excited state, if the energy of the photon matches the difference in energy between the two quantum mechanical levels according to the fundamental relation $\Delta E = h\nu$.

Because the absorption of light results from an interaction with the electric field of the incident light, the molecule, not surprisingly, retains a memory of the polarization of the incident light. The electrons in the molecule not only vibrate with the same frequency as the incident light, they also vibrate with the same polarization (i.e., they vibrate in the same direction as the electric field of the light). When the excited-state molecule fluoresces, the electric field of the emitted light is oriented in the same direction that the electrons were oscillating in the molecule. If the molecule did not rotate between the time it absorbs and the time it emits a photon, then the electric field of the incident light and the fluorescent light would be oriented in the same direction. In other words, the polarizations of the incident and fluorescent light would be the same.

Of course, excited molecules in solution do not remain stationary but rotate chaotically because of interactions (collisions) with the solvent molecules. Molecules remain in the excited state for a small but finite amount of time, and in general, we cannot assume the molecules do not rotate during this time. In this experiment, linearly polarized light irradiates a sample of chlorophyll molecules. If the lifetime of the excited state is long compared with the rotation speeds of the molecules in solution, the fluorescent light will be almost completely depolarized, that is, it will retain almost no memory of the polarization of the incident light. The fluorescent light can be expected to remain highly polarized only if the lifetime of the excited state is shorter than the time required for chlorophyll molecules to rotate significantly in solution or if it were possible to observe the fluorescence emitted just after excitation. Chlorophyll is a large, bulky molecule and thus tends to rotate slowly in solution. Accordingly, the extent of fluorescent depolarization for chlorophyll is relatively low at room temperature. Increasing the temperature increases the energy of the solvent molecules as well as the chlorophyll molecules, and on average the excited chlorophyll molecules rotate faster in solution, increasing the extent of depolarization.

In this experiment, the extent of fluorescent depolarization is measured indirectly by positioning the second polarizer to transmit light preferentially with the same polarization as the incident light. As the extent of depolarization increases, the intensity of light at the detector decreases. Little interference in these measurements would be expected from scattered laser light. As explained in the introduction to Chapter 2, almost all light scattered at right angles to the laser beam should have the same polarization as the incident light and be transmitted through the polarizer. Depolarization of scattered light is normally caused by multiple scattering and molecular anisotropy, and these factors do not depend on temperature.

Figure 5-4-2 shows a typical plot of light intensity at 90° vs. temperature. As ex-

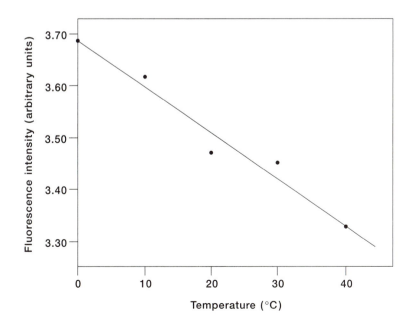

FIGURE 5-4-2. Fluorescence intensity viewed through polarizer as a function of temperature. Decreasing intensity indicates increasing depolarization.

pected, the fluorescent light is significantly more polarized at lower temperatures than at higher temperatures. The extent of depolarization increases approximately linearly with temperature (at least over the temperature range studied), which reflects the effect of increasing temperature to increase the fraction of molecules that have rotated appreciably before emitting.

MAGNETIC SUSCEPTIBILITY MEASUREMENTS

A laser pendulum apparatus is constructed and used to measure the magnetic properties of powdered crystalline samples.

DEGREE OF DIFFICULTY

Experimental: difficult
Conceptual: difficult

MATERIALS

- laser
- computer-interfaced light-detection system
- permanent magnet[1]
- 1-mL vials with plastic caps
- ring stand and clamps
- mortar and pestle
- various inorganic compounds (such as those listed in Table 5-5-1)
- thread
- epoxy

PROCEDURE

1. Pierce a plastic vial cap, insert one end of a 0.28-m thread into the cap, and affix the end to the cap, using epoxy. Attach a vial to the cap, and suspend the vial by the string so that it is free to swing. Align the permanent magnet (1-in. pole gap or smaller) as shown in Figure 5-5-1, and make sure that the sample sits in the upper quarter of the magnet's gap when at rest. Align the laser and the photodiode of the light-detection system with the hanging vial. A small drop of glue may be used on the thread to make a larger light block if the thread is too narrow to be "seen" by the photodiode.

2. Prepare test samples by finely grinding crystals of the various inorganic substances listed in Table 5-5-1 with a mortar and pestle. Add approximately 0.2 g of each to the vials. Carefully clean the mortar and pestle after grinding each sample to avoid contaminating the samples. Cap the vials after adding the samples.

---

1. The magnet we used had a 1-inch pole gap and a field strength of 5 kG. Magnets with these specifications are difficult to find commercially, however. The rare earth magnets that have recently become commercially available have very high field strengths and are economical to use. Two of them could be mounted approximately 1 inch apart to generate approximately the same magnetic field.

174

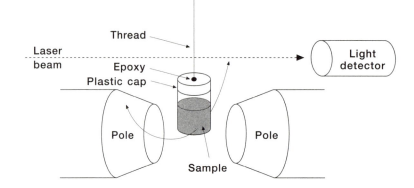

FIGURE 5-5-1. Schematic diagram of laser pendulum apparatus for measuring magnetic susceptibilities.

3. Test the laser-detection system using one of the samples. Once the positions of the magnet, laser, and photodiode are optimized, do not move them. To avoid interfering with the other equipment, change samples by snapping off the cap and snapping on a new vial.

4. Swing the sample through small angle displacements, and record the photodiode response on the computer to determine the period of the pendulum with each compound. Keep the method of swinging and the swing arc as uniform as possible for all samples.

TABLE 5-5-1. MAGNETIC SUSCEPTIBILITY RESULTS: EXPERIMENTAL vs. LITERATURE VALUES

Compound	μ_M Observed[a]	μ_M Literature[b]	n Calc[c]	n Theory[d]
$ZnCl_2$	0	0		
$SnCl_2$	0	0		
$CuBr_2$	1.81	1.28[e]		
$CuSO_4 \cdot 5H_2O$	1.88	1.88[e]		
$Ni(NO_3)_2 \cdot 6H_2O$	2.88	3.24[e]		
$CrCl_3 \cdot 6H_2O$	3.89	4.08[e]		
$CoCl_2 \cdot 6H_2O$	4.87	4.87[e]		
$Co(OAc)_2 \cdot 4H_2O$	4.99	4.98[e]		
$FeSO_4 \cdot 7H_2O$	6.22	5.22[e]		
$K_4Fe(CN)_6 \cdot 3H_2O$	0	0[e]	0	0
$K_3Fe(CN)_6$	2.69	2.4[e]	1.87	1
$Ni(NH_3)_6Br_2$	3.20	3.24[f]	2.35	2
$K_3Cr(C_2O_4)_3 3H_2O$	3.80	3.75[g]	2.92	3
$Fe(NH_4)_2(SO_4)_2 \cdot 6H_2O$	5.49	5.25[h]	4.58	4

[a] Total magnetic moment in Bohr magnetons calculated as described in discussion section.
[b] Calculated from χ_M values with the assumption that the magnetic polarizability is negligible.
[c] Calculated from μ_M (observed) with the assumption that the orbital contribution to μ_M is negligible.
[d] Number of unpaired electrons according to reference 4.
[e] According to reference 5.
[f] According to reference 6.
[g] According to reference 7.
[h] According to reference 8.

HAZARDS AND PRECAUTIONS

Tin (II) chloride is corrosive and skin absorbable. Nickel (II) nitrate hexahydrate is a strong oxidizer. Most chromium salts are carcinogens. Many of the other substances are toxic or suspected carcinogens. Review the Material Safety Data Sheets for each compound used. Wear gloves when handling the compounds, and do all sample grinding in a fume hood. Wear eye protection at all times.

DISPOSAL

Chromium salts and some cyano-containing compounds require special disposal procedures. All other substances can be flushed from the glass vials with water and disposed of in an aqueous waste container. Follow established procedures for handling and disposal of aqueous wastes.

DISCUSSION

The magnetic properties of a compound provide clues to its electronic structure particularly whether or not all of its electrons are paired. A fundamental measurement of magnetic behavior is the magnetic susceptibility of a substance, which measures how much its magnetic moment changes in the presence of a magnetic field. In this experiment, the origins and implications of this effect are examined. The laser-pendulum apparatus used here replaces the more costly and complex Gouy or Faraday balances traditionally used for measurement of the effects of an applied magnetic field (1).

One important result of classical electromagnetism is that spinning and rotating charges have associated magnetic moments. Many molecules have small magnetic moments (μ_M) that arise from two sources: the magnetic moments of the electrons themselves (μ_{spin}) and the magnetic moment that arises from the orbital motion of the electrons around the nuclei (μ_{orb}):

$$\mu_M = \mu_{spin} + \mu_{orb} \tag{5-5-1}$$

The magnetic moment that arises from the orbital motion of an electron depends on the particular orbital the electron occupies and cannot be computed easily. For the substances used in this experiment, however, the orbital contribution to the magnetic moment is far smaller than the contribution of the electron spin, and as a rough approximation it may be assumed that the magnetic moment arises solely from the spins of the unpaired electrons; that is,

$$\mu_M \approx \mu_{spin} \tag{5-5-2}$$

When this relationship holds true, the magnetic moment is sometimes called a *spin-only magnetic moment*.

The magnetic moment of an electron arises from its intrinsic spin of $s = 1/2$ and, by quantum mechanics, is given to be

$$\mu_{one\ e^-} = 2\mu_B\sqrt{s(s+1)} \tag{5-5-3}$$

where μ_B is a constant known as the Bohr magneton and $s = 1/2$. Electrons with opposite spins have magnetic moments oriented in opposite directions, and the magnetic moments of any paired electrons cancel each other. Substances with unpaired electrons have permanent magnetic moments that arise from electron spins. If we define S, the total spin quantum number, as the sum of all the spins in an atom or molecule, then the total magnetic moment is

$$\mu_{spin} = 2\mu_B \sqrt{S(S+1)} \tag{5-5-4}$$

Since the spin on the electron is 1/2, the number of unpaired electrons (n) is related to S by

$$S = n/2 \tag{5-5-5}$$

and equation 5-5-4 can be rewritten as

$$\mu_{spin} = \mu_B \sqrt{n(n+2)} \tag{5-5-6}$$

Thus, a measurement of the value of μ_M can be used to deduce n, the number of unpaired electrons in a molecule, provided that μ_{orb} is negligible.

The value of the magnetic moment of a substance can be determined indirectly by observing how the substance interacts with a magnetic field. When any substance is placed in a permanent magnetic field, the field induces a magnetic moment in the object that depends largely on the magnetic moment of the substance. The strength of the induced magnetic field (B) can be related to the strength of the external magnetic field (H) according to the relationship

$$B = H(1 + 4\pi X) \tag{5-5-7}$$

where X is the magnetic susceptibility per unit volume, a dimensionless quantity that represents the intensity of magnetization per unit field strength. A more useful quantity than X for performing calculations is the molar magnetic susceptibility X_M, which is defined as the magnetic susceptibility times the molar volume of the substance:

$$X_M = X\left(\frac{M}{\rho}\right) \tag{5-5-8}$$

where M is the molecular weight (g/mol) of the substance, ρ is the density (g/cm^3), and X_M has units of cm^3/mol.

The molar magnetic susceptibility is related to the magnetic moment of a particular substance by

$$X_M = N\left(\alpha_M + \frac{\mu_M^2}{3kT}\right) \tag{5-5-9}$$

where N is Avogadro's number, k is Boltzmann's constant, T is temperature, α_M is the magnetic polarizability, and μ_M is the magnetic moment. The magnetic polarizability is an inherent property of substances and is generally small and negative, whereas the value of the magnetic moment is large and positive. Thus, substances without a permanent magnetic moment have a value of X_M that is small and negative, which means that the magnetic moment induced in the substance points in a direction opposite to the external field and acts to cancel the field strength inside the object. These substances are called *diamagnetic*. For substances that have a permanent magnetic moment, the positive $(\mu_M^2/3kT)$ term far outweighs the negative α_M term at room temperature. These substances are called *paramagnetic,* and the induced magnetic field points in a direction parallel to the external field and acts to enhance the magnetic field in the substance.

For many paramagnetic substances, the α_M term is so much smaller than the $(\mu_M^2/3kT)$ term as to be negligible. In this case, equation 5-5-9 can be rewritten as

$$\mu_M = \sqrt{\frac{3kT\chi_M}{N}} \tag{5-5-10}$$

By evaluating the constants explicitly, we obtain

$$\mu_M = 2.84\sqrt{\chi_M T} \tag{5-5-11}$$

in units of Bohr magnetons. Thus, for the substances studied in this experiment, the value of χ_M directly reflects the number of unpaired electrons (2,3).

The laser pendulum system used in this experiment allows for a remarkably simple determination of the molar magnetic susceptibility. In the absence of an external field, the period of a simple pendulum (τ) can be approximated by

$$\tau = 2\pi\sqrt{\frac{\ell}{g}} \tag{5-5-12}$$

where ℓ is the length of the string and g the gravitational constant. When a paramagnetic sample swings back and forth in an inhomogeneous magnetic field, an additional force is developed that is proportional to χH^2, and this force always acts to pull the pendulum into the region of highest magnetic field strength. This extra force acts in the same direction as the gravitational force on the sample, and the pendulum behaves as if the value of g had increased to a value g'. Specifically, the period of the pendulum is shortened and the new period τ' is given by

$$\tau' = 2\pi\sqrt{\frac{\ell}{g'}} \tag{5-5-13}$$

Solving equations 5-5-12 and 5-5-13 for g and g' gives

$$g = \frac{4\pi^2\ell}{\tau^2} \tag{5-5-14}$$

and

$$g' = \frac{4\pi^2\ell}{\tau'^2} \tag{5-5-15}$$

Subtracting these two equations yields

$$\Delta g = g' - g = 4\pi^2\ell\left(\frac{1}{\tau'^2} - \frac{1}{\tau^2}\right) \tag{5-5-16}$$

The value of Δg multiplied by the mass m of the sample is equivalent to an effective weight change of the sample

$$\Delta W = m\,\Delta g \tag{5-5-17}$$

As mentioned previously, the effective weight change should be proportional to χ

$$\chi \propto \Delta W \tag{5-5-18}$$

implying that

$$\chi_M \propto \Delta W \left(\frac{M}{\rho}\right) \tag{5-5-19}$$

The procedure used to determine the magnetic moment of an unknown sample is to measure the period of the pendulum with and without the sample, calculate the value of ΔW, and then compare the ΔW to a standard with a known molar magnetic susceptibility. For this experiment, a useful standard is ($CoCl_2 \cdot 6H_2O$), whose molar magnetic susceptibility $(\chi_M)_s$ is 9710.0×10^{-6} cm^3/mol. Using formulas 5-5-19 and 5-5-11, the magnetic moment of an unknown sample can be related to the standard reference sample by

$$\mu_M = 2.84 \left[\left(\frac{\Delta W}{\Delta W_s}\right)\left(\frac{\rho_s}{\rho}\right)\left(\frac{M_s}{M}\right)(\chi_M)_s T\right]^{1/2} \tag{5-5-20}$$

where the subscript S refers to the standard, and the magnetic polarizability of the sample is assumed to be negligible.

To summarize, the laser and computer-interfaced light-detection system can be used to make sensitive measurements of the period of the pendulum. The difference between the period of the pendulum with and without the sample is used in the calculation of an effective weight change (equations 5-5-16 and 5-5-17). This effective weight change can be compared with the weight change of a standard with a known magnetic susceptibility, and can thus be used in the calculation of the magnetic moment of the sample (equation 5-5-20). Assuming that the magnetic moment of the sample arises from electron spin alone, the effective number of unpaired electrons can be calculated according to equation 5-5-6.

Table 5-5-1 compares the experimental values of μ_M determined for 14 compounds with their literature values. The accuracy achieved using this technique was less than that reported using the standard Gouy balance method, but with a little practice and experimental optimization, accuracies of up to 95% can be achieved. For the final five octahedral complexes, the table also shows calculated and theoretical values for the number of unpaired electrons n. The reasonably large discrepancies between the calculated and theoretical values arise partly from experimental error but also indicate that these compounds have significant orbital contributions to their magnetic moments.

REFERENCES

1. Spencer, B., Zare, R.N. *J. Chem. Ed.* 65 (1988) 277.
2. Cotton, F.A.,Wilkenson, G. *Advanced Inorganic Chemistry,* 3rd ed. Interscience; New York: 1972; 540–541.
3. McQuarrie, D.A., Rock, P.A. *General Chemistry,* 3rd ed. W.H. Freeman; New York: 1991; Chapter 28: 993–1017.
4. Figgis, B.N. *Introduction to Ligand Fields.* Interscience; New York: 1966; Chapter 10: 248–292.
5. Weast, R.C., ed. *CRC Handbook of Chemistry and Physics,* 67th ed. CRC; Boca Raton, FL: 1986; E-119–124.
6. Bose, D.M. *Z. Physik* 65 (1930) 677.
7. Johnson, C.H. *Trans Faraday Soc.* 38 (1932) 845.
8. Kirschner, S., Albinak, M.J., Bergman, J.G. *J. Chem. Ed.* 39 (1962) 576.

CHAPTER 6

PHOTOCHEMISTRY

Photochemistry is the study of how matter can be stimulated by light to undergo chemical change. This introduction considers the salient features of photochemical reactions, and the experiments provide examples of how lasers can be used to initiate and monitor photochemical reactions. An understanding of the section titled "Molecular Electronic Structure" in the introduction to Chapter 5 is helpful for understanding many of the topics in this chapter.

The distinguishing feature of photochemical reactions is that at least one of the chemical species involved absorbs light, which implies that excited-state molecules participate in the reaction. In each section below, we consider some interesting differences between photochemical reactions and nonphotochemical reactions, as well as the potential advantages of using light to carry out chemical change. A more complete discussion appears in a series of photochemistry experiments designed for high school and undergraduate use (1).

THERMODYNAMIC ADVANTAGES OF ACCESSING EXCITED STATES

In a photochemical reaction, the energy absorbed by the reactants can help to drive the reaction. Consider a simple reversible unimolecular reaction, such as an isomerization, which can be represented as

$$A \longleftrightarrow B \qquad (6\text{-}1)$$

If the mixture of A and B is in equilibrium and the A to B transition is uphill (i.e., it consumes free energy), only small amounts of product B will be produced. In this case, using a photochemical pathway could be a valid method of increasing the concentration of product B. By absorbing light, the reactant A might be able to enter into an excited state that has a higher energy than the

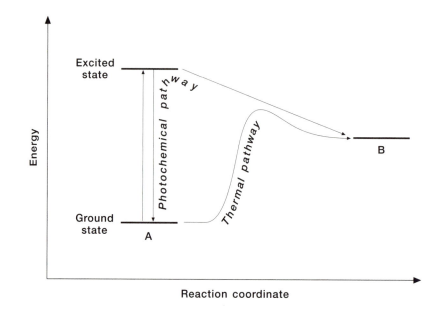

FIGURE 6-1. Comparison of photochemical and thermal (nonphotochemical) reaction pathways.

product, as shown in Figure 6-1. If some mechanism exists by which the excited-state reactant can decay to the product, then product B can be made in much greater quantities than by nonphotochemical means. One potential difficulty with the photochemical scheme is that any product B formed could revert to the reactant because it is thermodynamically unfavorable compared with the reactant A. If a substantial activation energy exists for the reverse reaction, however, then the reversion of the product to the reactant will be slow, especially at low temperatures.

SELECTIVITY

Photochemical reactions can often be designed so that only certain chemical species absorb light and enter into excited states. For instance, in the unimolecular reaction considered above, the substances A and B could have different absorption spectra, and so a wavelength of light could be chosen that excites A but not B. The ability to add energy to some substances but not others gives photochemistry an extra degree of control over thermal reactions. By contrast, heating a reaction mixture adds energy equally to reactants and products. Lasers are particularly useful for selectively exciting molecules because they emit light of only one wavelength.

PROPERTIES OF EXCITED-STATE VS. GROUND-STATE MOLECULES

Molecules in excited states generally have different electron distributions than they do in the ground state and thus are able to react differently. Excited-state molecules often accept and donate electrons more readily than they do in the ground state. Figure 6-2 provides a simple way of conceptualizing these differences. The ionization energy of a molecule is a rough measure of its electron-donating ability. As pictured in Figure 6-2a, excited states of molecules have lower ionization energies and thus are better electron donors *(reducing agents)*. Conversely, the electron affinity of a molecule is a rough measure of its electron-accepting ability. Excited states often have larger electron affinities, as diagrammed in Figure 6-2b, and thus are expected to be better electron acceptors *(oxidizing agents)*.

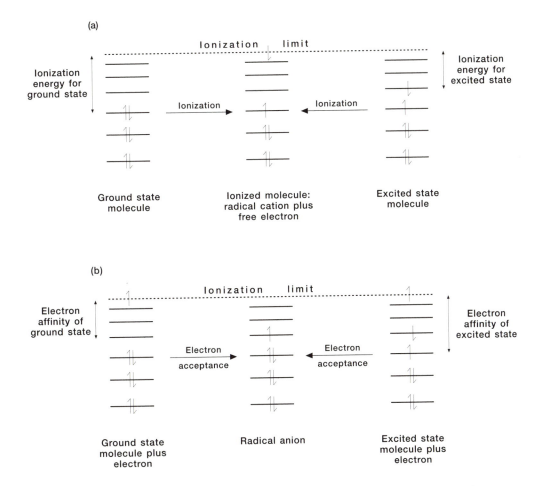

FIGURE 6-2. Comparison of the ionization (panel **a**) and electron acceptance (panel **b**) energies of ground- and excited-state molecules.

Because of their different electron-transfer capabilities, excited states of molecules generally have very different redox properties than ground-state molecules. The electron-transfer capabilities of excited-state molecules are often so enhanced that many photochemical reactions in solution begin with a complete transfer of one electron to or from an excited-state molecule and proceed by a free-radical mechanism.

In the experiments that follow, we present some examples of photochemical reactions and demonstrate how lasers can be used to initiate and monitor these reactions. Lasers are particularly useful light sources for photochemical reactions because the intensity of the laser beam permits faster reaction times and the monochromaticity of the laser allows the possibility of selectively exciting one or more of the reactants. Experiments 6-5 and 6-6 also demonstrate how lasers can be useful for gathering kinetic information by indirectly monitoring the concentration of a reactant or product.

REFERENCE

1. Tokumaru, K., Coyle, J.D. *Pure and Applied Chemistry.* 64 (1992) 1343.

PHOTOBLEACHING OF METHYLENE BLUE

A sample of methylene blue and triethylamine is irradiated with a red laser, which leaves a bleached channel where the laser passed through the solution. This photobleaching provides evidence of a triplet-state methylene blue intermediate that has enhanced electron-acceptor capabilities. It is this experiment that is pictured on the cover of this book.

DEGREE OF DIFFICULTY

Experimental: easy
Conceptual: moderate

MATERIALS

- red He–Ne laser[1]
- screw-cap vial, 10 mL
- 3 vials with caps, 1 mL
- pipettes
- 2 mg of methylene blue (Sigma product #MB-1. Approx. $12 for 25 g)
- 5 drops of triethylamine (Sigma product #T0886. Approx. $6 for 100 mL)
- green laser or other intense green light source (if available)

PROCEDURE

1. Mix 10 mL of water, 1–2 mg of methylene blue, and 5 drops of triethylamine in a 10 mL screw-cap vial.

2. Pipette 0.75-mL portions of this solution into each of three 1-mL vials. Cap the vials.

3. Place the remaining liquid in the 10-mL vial in bright sunlight for approximately 5 minutes, and observe any color change in the solution. Remove the sample from the sunlight and observe any color change.

4. Place the three 1-mL vials in a dark room. Irradiate one with a red laser and one with a green laser or any other reasonably intense green light source (if available). The third vial serves as a control. Monitor the vials occasionally over a 1-day period.

HAZARDS AND PRECAUTIONS

Triethylamine is corrosive and easily permeates the skin. Prepare the solutions under a fume hood, and wear gloves when handling the solutions. Wear eye protection at all times.

---

1. For this experiment, the laser needs to have an output wavelength near 630 nm. Red diode lasers with outputs near 685 nm may not be useful because the methylene blue does not absorb strongly at this wavelength.

DISPOSAL

Collect waste solutions in an aqueous waste bottle and dispose of them in an approved manner.

DISCUSSION

Methylene blue ($^1MB_o{}^+$) is an organic dye (Figure 6-1-1) that absorbs strongly in the

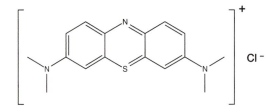

FIGURE 6-1-1. Molecular structure of methylene blue.

red region of the visible spectrum (Figure 6-1-2). When a methylene blue molecule absorbs red light, it makes a transition to an excited singlet state, denoted by $^1MB^+$. The excited singlet state of methylene blue either returns to the ground state or converts

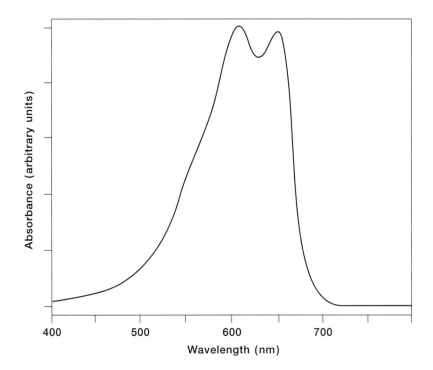

FIGURE 6-1-2. Absorption spectrum of methylene blue in methanol.

to an excited triplet state.[2] The triplet state, which is represented below as $^3MB^+$, has a reasonably long lifetime ($4.2 \times 10^{-5}s$), and for reasons described in the introduction to this chapter, the excited triplet state is a better electron acceptor (oxidizing agent) than the ground-state form.

The photobleaching observed in this experiment is the result of a photochemical redox process. Ground-state methylene blue does not react noticeably with triethylamine, but the excited triplet state of methylene blue efficiently oxidizes the triethylamine. In the process, the methylene blue reduces to a radical, which is colorless. The photobleaching process can be summarized as follows:

$$^1MB_0 + \xrightarrow{h\upsilon} {}^1MB^+ \rightarrow {}^3MB^+$$
$$\text{(blue)}$$
$$^3MB^+ + :NR_3 \rightarrow MB\cdot + \cdot NR_3^+$$
$$\text{(clear)}$$

(6-1-1)

where $:NR_3$ represents triethylamine or any similar amine.

Sunlight can be used to promote this reaction, but the use of a red He–Ne laser (632.8 nm) provides a more striking demonstration of the photobleaching: noticeable bleaching in the bulk solution and a completely bleached channel where the laser beam passed through the solution. (See cover illustration of this book.) The control sample, which is not irradiated, displays no noticeable bleaching, which confirms that bleaching results from a photochemical process and that the excited state of methylene blue is a stronger oxidizing agent than the ground state. The sample irradiated with green light also shows no evidence of bleaching because methylene blue does not absorb green light as strongly as it absorbs red light.

Upon standing in the shade, the photochemically bleached solution rapidly regains its initial blue color. Note that shaking the colorless bleached solution increases the speed of the reversion to blue and that, even under full sunlight, the solution never becomes totally blue at the surface of the liquid. Can you think of an explanation for this observation?

REFERENCE

1. Lutoshkin, V.I., Volkon, S.V. *Theor. Exp. Chem.* 13 (1977) 278.

2. Strictly speaking, the conversion of the singlet to the triplet state may not be an intersystem crossing of the isolated molecule because the conversion mechanism may involve interactions with nearby ground-state methylene blue molecules (1).

PHOTOISOMERIZATION OF DIMETHYLMALEATE

Under catalysis by bromine, *cis*-dimethylmaleate isomerizes to *trans*-dimethylmaleate under irradiation by blue light but not by red or green light.

DEGREE OF DIFFICULTY

Experimental: easy
Conceptual: moderate

MATERIALS

- red laser
- green laser or high-intensity lamp with green filter
- high-intensity lamp with blue filter
- lens suitable for defocusing laser beam
- 3 vials with tight-fitting caps
- 2.16 g of *cis*-dimethylmaleate (dimethyl ester of maleic acid) (Aldrich product #23,819-8. Approx. $9 for 5 g)
- 10 mL of carbon tetrachloride
- 1 mL of 0.6-M bromine in carbon tetrachloride
- 1 mL of 0.6-M iodine in carbon tetrachloride *(optional)*

PROCEDURE

1. Prepare a 1.5-M solution of *cis*-dimethylmaleate (MW = 144.13) by dissolving 2.16 g in 10 mL of carbon tetrachloride.

2. Add 6 drops of a solution that is 0.6-M bromine in carbon tetrachloride, stir quickly, and transfer a small amount of this solution into 5 vials.

3. Irradiate each of 4 vials with one of the following:
   a. Defocused red laser (defocus the laser beam by placing the lens in the path of the laser beam and adjusting the position of the lens so that the laser light expands to irradiate a large portion of the surface of the vial)

   b. Defocused green laser (or other green light source)

   c. High-intensity lamp with blue filter

   d. Bright sunlight

   Place the fifth vial in the dark to act as a control. Performing the illuminations in a dark room is best because ordinary room light may be sufficient to cause isomerization.

4. *(Optional)* Repeat steps 2–4 using iodine instead of bromine.

HAZARDS AND PRECAUTIONS

Carbon tetrachloride is a carcinogen. Bromine and iodine are strong oxidizers and are corrosive. Work in a fume hood and wear gloves when handling solutions. Wear eye protection at all times.

DISPOSAL

Bromine is classified as an extremely hazardous waste. Because of the corrosive nature of bromine and iodine, collect any solutions containing these compounds in separate, carefully labeled waste containers, and follow appropriate disposal procedures. Chlorinated solvents, such as carbon tetrachloride, require special disposal procedures.

DISCUSSION

The reaction studied in this experiment is the isomerization reaction shown in Figure 6-2-1. The reactant is the dimethyl ester of maleic acid (*cis*-dimethylmaleate), and the

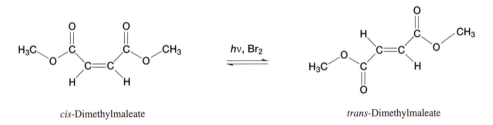

cis-Dimethylmaleate   trans-Dimethylmaleate

FIGURE 6-2-1. Isomerization reaction of *cis*-dimethylmaleate to *trans*-dimethylmaleate.

product, which is insoluble under the reaction conditions, is the corresponding ester of fumaric acid (*trans*-dimethylmaleate). Bromine acts as a catalyst. The reaction proceeds very slowly in the dark, but in strong sunlight, crystalline *trans*-dimethylmaleate appears within minutes.

The most likely mechanism for this photochemical isomerization is shown in Figure 6-2-2. Often, diatomic molecules have lower bond strengths in excited states than in the ground state because the absorption of light promotes an electron in a bonding molecular orbital to a higher energy excited molecular orbital that may destabilize the chemical bond. In this experiment, bromine absorbs visible light, and in the excited state, the Br–Br bond breaks more readily to produce two bromine radicals. The bromine radicals then react with the *cis*-dimethylmaleate, thereby producing a brominated hydrocarbon radical (compound A in Figure 6-2-2). Because this radical does not have a double bond, the substituents are free to rotate to a different conformation (compound B). If the bromine radical dissociates from the organic molecule in this new conformation, then the double bond is reformed, except that the ester groups are in *trans* positions to each other (*trans*-dimethylmaleate).

This mechanism is supported by the fact that crystalline product forms rapidly under irradiation by blue light but not by red or green light. *cis*-Dimethylmaleate is colorless,

188

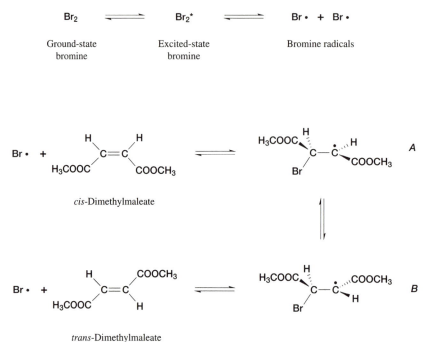

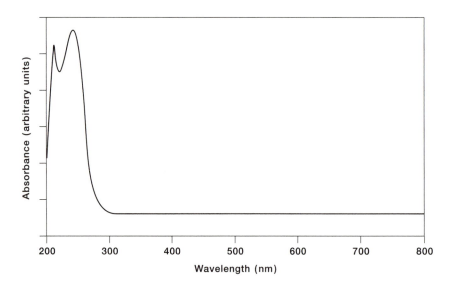

FIGURE 6-2-2. Likely mechanism for the reaction depicted in Figure 6-2-1. Compound A is the brominated hydrocarbon radical, and compound B is a different rotational conformation of compound A.

indicating that it does not absorb strongly in the visible region of the spectrum (Figure 6-2-3). As a result, the reactant is not likely to be responsible for the wavelength dependency of the extent of the reaction. On the other hand, bromine strongly absorbs blue light but not red light (Figure 6-2-4), which supports the assertion that bromine is the initiator of this photochemical reaction.

FIGURE 6-2-3. Absorption spectrum of *cis*-dimethylmaleate in methanol.

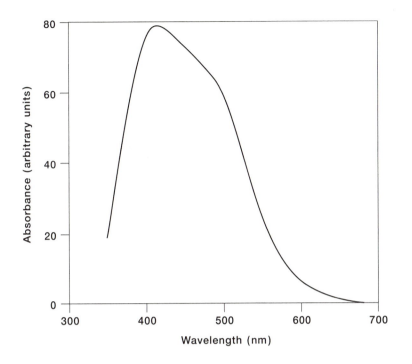

FIGURE 6-2-4. Absorption spectrum of bromine in carbon tetrachloride.

Iodine absorbs green light strongly (Figure 6-2-5), and iodine might be expected to catalyze the isomerization in much the same manner as bromine. However, samples prepared in the same manner as for bromine and exposed to sunlight give no indication of isomerization after several hours. Product is observed in the vials only after several days of exposure.

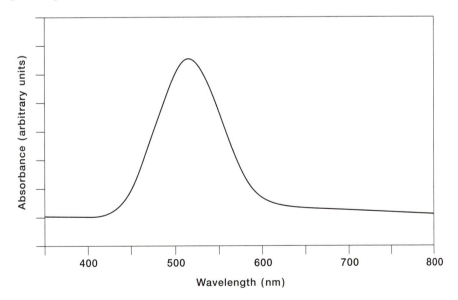

FIGURE 6-2-5. Absorption spectrum of iodine in chloroform.

PHOTOINDUCED POLYMERIZATION OF ACRYLAMIDE

A laser is used as the initiator for the photoinduced polymerization of acrylamide.

DEGREE OF DIFFICULTY

Experimental: easy
Conceptual: moderate

MATERIALS

- red He–Ne laser[1]
- 2 microscope slides
- 150-mL Erlenmeyer flask
- 100-mL volumetric flask
- 15 g of gelatin
- 15.57 g of acrylamide
- 3.3 mL of triethanolamine
- 0.025 mg of methylene blue

PROCEDURE

1. Prepare a 15% gelatin solution by dissolving 15 g of gelatin in a small amount of hot water in a 150-mL Erlenmeyer flask. Once the gelatin is completely dissolved, add enough hot water to bring the total volume in the flask to approximately 100 mL. Place a magnetic stir bar in the flask, and stir with slight warming until ready for use.

2. In a fume hood, add 15.57 g of acrylamide, 0.025 g of methylene blue, and 3.3 mL of triethanolamine to a 100-mL volumetric flask. Fill the flask with the gelatin solution from step 1 to the 100-mL mark and mix well.

3. In a fume hood, place a single drop of the liquid on each of two microscope slides. Still under the fume hood, pass the beam of the laser through one of the drops, parallel to the microscope slide. Keep the second slide as a control not exposed to laser light. Allow the reaction to proceed for 24 hours, preferably in the dark.

4. Compare the appearance of the two drops.

----

1. For this experiment, the laser needs to have an output wavelength near 630 nm. Red diode lasers with output near 685 nm may not be useful because the methylene blue does not absorb strongly at this wavelength.

HAZARDS AND PRECAUTIONS

Acrylamide is a NIOSH carcinogen. Always handle acrylamide under a fume hood. Triethanolamine is corrosive. Wear eye protection at all times. Wear gloves when handling solutions or solids that contain acrylamide or triethanolamine.

DISPOSAL

Collect waste solutions in an organic waste container, and dispose of the container in an approved manner. Label all waste containers fully as to their contents.

DISCUSSION

Polymers are large organic molecules composed of structural units called monomers linked together in long chainlike structures. Examples of polymers include plastics, many types of fibers, and cellulose. Many common synthetic polymers are vinyl polymers composed of monomers derived from ethylene. The polyacrylamide prepared in this experiment is one such polymer. The structures of the acrylamide monomer and polyacrylamide are shown in Figure 6-3-1.

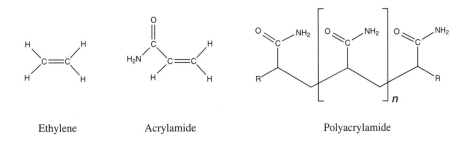

Ethylene   Acrylamide   Polyacrylamide

FIGURE 6-3-1. Molecular structures of ethylene, acrylamide, and polyacrylamide.

The process by which monomers become linked into polymer chains is called *polymerization*. The polymerization process observed in this experiment can be represented as

$$n \text{ Acrylamide} \longrightarrow (\text{acrylamide})_n \equiv \text{polyacrylamide} \qquad (6\text{-}3\text{-}1)$$

The value of n is typically on the order of 1,000. This polymerization is one of many polymerization reactions that proceed by free-radical mechanisms. Figure 6-3-2 outlines the free-radical polymerization of ethylene monomers. In the initiation step, a radical reacts with a monomer and forms a new radical. This radical can then react with another monomer, thereby forming a radical composed of two monomers. This process continues and the polymer radical grows until the chain reaction is terminated, usually by the reaction of two radicals with each other.

Several techniques are available to generate radicals in solution to initiate the free-radical polymerization. This experiment provides an example of how light can be used to initiate a polymerization reaction. A likely mechanism (1) for the initiation process is

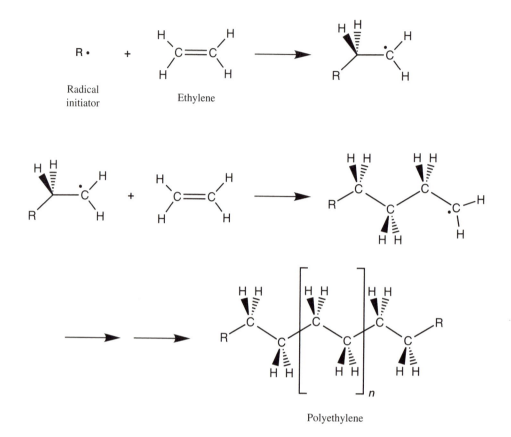

FIGURE 6-3-2. Free-radical polymerization of ethylene monomers.

shown in Figure 6-3-3. As discussed in experiment 6-1, methylene blue can be excited indirectly to a triplet state by absorption of red light. This triplet state of methylene blue, which is a stronger oxidizer than the ground state, then participates in a redox reaction, in which it abstracts a hydrogen from one of the alcohol groups on triethano-lamine. The result is two radicals that may react with other species in solution to form new radicals. Which radical actually initiates the polymerization of acrylamide remains unclear. Diatomic oxygen most likely plays a role in regenerating ground-state methyl-ene blue (1).

The polymerization reaction studied in this experiment is carried out in a gelatin solution and requires approximately 24 hours of irradiation. The polymeric areas, which are clearer and harder than the original solution, appear only where the red laser light has propagated through the gel.

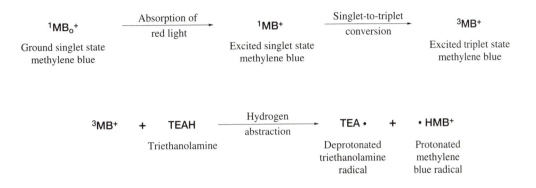

FIGURE 6-3-3. Probable initiation steps in the photoinduced polymerization of acrylamide in the presence of methylene blue and triethanolamine.

REFERENCE

1. Lutoshkin, V.I., Volkov, S.V. *Theor. Exp. Chem.* 13 (1977) 278.

PHOTOOXIDATION OF DIPHENYLISOBENZOFURAN

The oxidation of diphenylisobenzofuran (DPBF) in the presence of methylene blue is initiated with red laser light, and the extent of the reaction is monitored with a spectrophotometer.

DEGREE OF DIFFICULTY

Experimental: moderate
Conceptual: moderate

MATERIALS

- red He–Ne laser[1]
- visible spectrophotometer
- cuvette, 1 cm[2]
- double convex lens
- mirror
- 4 volumetric flasks, 25 mL
- 400 mL of methanol
- 100 mg of diphenylisobenzofuran (Sigma product #D5016. Approx. $8 for 500 mg)
- 100 mg of methylene blue (Sigma product #MB-1. Approx. $12 for 25 g)
- pipettes

PROCEDURE

1. Set up the apparatus as shown in Figure 6-4-1. Without the lens in place, position the laser beam to pass through the center of the cuvette. Position the lens so that the laser beam expands to uniformly fill the bottom of the cuvette space. Monitor all reactions in a dimly lit or dark room, because ordinary room light is sufficient to stimulate the reaction.

2. Prepare 25 mL of the following solutions:

   a. 2×10^{-4} M diphenylisobenzofuran (MW = 270.3) in methanol

   b. 1.6×10^{-6} M methylene blue (MW = 373.9) in methanol

3. Determine the wavelength of maximum absorbance for DPBF by recording the absorbance spectrum of solution *a* from 350 nm to 750 nm. Use methanol in the reference cell. The peak absorbance should be near 420 nm, and the peak absorption wavelength should be used to monitor the reactions (see step 4).

---

1. For this experiment, the laser needs to have an output wavelength near 630 nm. Red diode lasers with outputs near 685 nm may not be useful because the methylene blue does not absorb strongly at this wavelength.

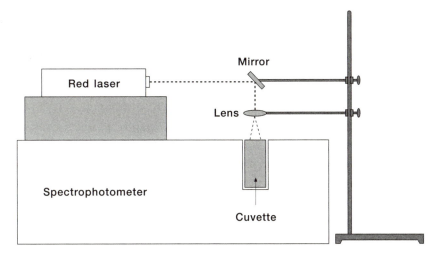

FIGURE 6-4-1. Schematic diagram of experimental setup for monitoring the photooxidation of diphenylisobenzofuran.

4. Using a pipette, transfer 0.7 mL of solution *a* and 0.7 mL of solution *b* into a cuvette. Mix quickly and place the cuvette in the spectrophotometer. Immediately begin irradiating the sample with the laser and recording the absorbance at the wavelength of maximum absorbance determined in step 3. Record the absorbance at least once a minute until the absorbance change is small (typically ~ 60 min).

5. *(Optional)* Repeat step 4, substituting methylene blue solutions of different concentrations. More concentrated methylene blue solutions should increase the rate of reaction.

HAZARDS AND PRECAUTIONS

Use caution when handling fused aromatic systems, such as diphenylisobenzofuran. Methanol easily permeates the skin. Use gloves when handling all solutions. Wear eye protection at all times.

DISPOSAL

Discard all solutions in an organic waste disposal bottle.

DISCUSSION

The reaction observed in this experiment is the oxidation of DPBF to *o*-dibenzoylbenzene in the presence of methylene blue. The reaction proceeds slowly in the dark; light is necessary for it to proceed at a reasonable rate. The most likely mechanism for the reaction (1,2) is depicted in Figure 6-4-2. As described in experiment 6-1, methylene blue strongly absorbs red light, which creates highly reactive triplet-state molecules. Oxygen in its ground state is a triplet-state molecule, and triplet-state oxygen dissolved

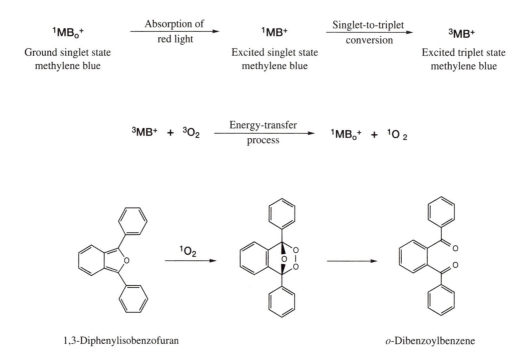

$$^1MB_o^+ \xrightarrow[\text{red light}]{\text{Absorption of}} {}^1MB^+ \xrightarrow[\text{conversion}]{\text{Singlet-to-triplet}} {}^3MB^+$$

Ground singlet state   Excited singlet state   Excited triplet state
methylene blue   methylene blue   methylene blue

$$^3MB^+ + {}^3O_2 \xrightarrow[\text{process}]{\text{Energy-transfer}} {}^1MB_o^+ + {}^1O_2$$

1,3-Diphenylisobenzofuran o-Dibenzoylbenzene

FIGURE 6-4-2. Probable mechanism for the photooxidation of diphenylisobenzofuran in the presence of methylene blue.

in solution reacts with the triplet-state methylene blue. The methylene blue returns to its singlet ground state, transferring its energy to the oxygen, which enters an excited singlet state. This process is called *triplet-triplet annihilation.* The excited singlet state of oxygen is a powerful oxidizing agent that reacts with the DPBF to form an unstable peroxide species, which then decays to the yellowish product, o-dibenzoylbenzene. Note that the overall reaction involves oxygen and DPBF but that the energy needed to drive the reaction comes from the red light absorbed by methylene blue!

The extent of the reaction can be monitored using a spectrophotometer because the reactant, DPBF, is yellow, and the product, o-dibenzoylbenzene, is colorless. As the reaction progresses, the absorbance of the solution at the wavelength of maximum absorbance of the reactant (~ 420 nm) decreases. Note that methylene blue absorbs only weakly at 420 nm (Figure 6-1-2) and thus does not interfere with these measurements. Figure 6-4-3 shows a plot of the observed disappearance of DPBF using a methylene blue concentration of 8×10^{-7} M. The curve is very nearly an exponential decay (Figure 6-4-4), which is an indicator of a first-order reaction.[2]

---

2. Justifying why the mechanism depicted in Figure 6-4-2 would produce first-order behavior is difficult, and in fact, we did not observe first-order behavior at higher methylene blue concentrations. Evidently, the reaction is not rigorously first order but only appears so under the specific conditions of this experiment (i.e., low methylene blue concentration).

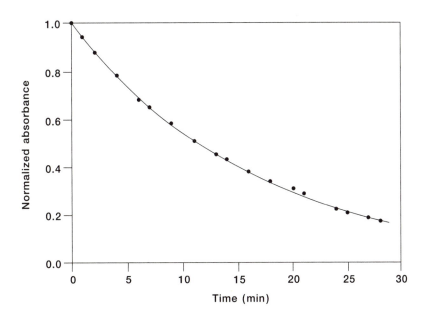

FIGURE 6-4-3. Kinetic data for the photooxidation of diphenylisobenzofuran using a methylene blue concentration of 8×10^{-7} M.

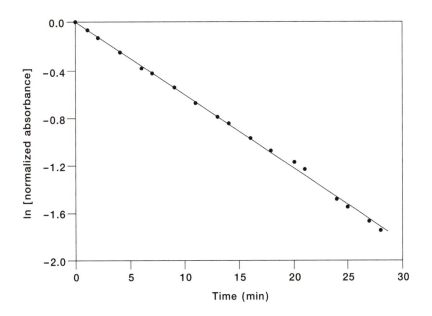

FIGURE 6-4-4. Natural logarithm of the normalized absorbance vs. time for the data depicted in Figure 6-4-3.

A quantitative study of the kinetics of this reaction would be challenging for several reasons. First, the oxygen concentration would need to be controlled somehow, such as by continuous aeration to ensure a constant concentration of oxygen. Second, the mechanism shown in Figure 6-4-2 is difficult to model kinetically because numerous species are involved and because the excited-state species have limited lifetimes. Further study of this reaction using different concentrations of methylene blue and controlling the oxygen concentration could yield insight into its kinetics. Note that DPBF is frequently used in studies of singlet oxygen because it is readily oxidized, and all of the steps involved in this reaction have been studied elsewhere. For example, Furve and Russell (3) report that the rate constant for the reaction of singlet oxygen with DPBF is $5.3 \times 10^8 \ M^{-1}s^{-1}$ and that the lifetime of singlet oxygen is on the order of microseconds. Foote (4) reports that the rate constant for the reaction of triplet-state methylene blue with ground-state oxygen is in the range of $1–3 \times 10^9 \ M^{-1}s^{-1}$.

REFERENCES

1. Wasserman, H.H., Murray, R.W. *Singlet Oxygen.* Academic; San Francisco: 1979; 435.
2. Young, R.H., Brewer, D.R. In Rånby, B., Rabek, J.F., eds. *Singlet Oxygen: Reactions with Organic Compounds and Polymers.* John Wiley and Sons; New York: 1978; 36.
3. Furve, H., Russell, K.E. In Rånby B., Rabek, J.F., eds. *Singlet Oxygen: Reactions with Organic Compounds and Polymers.* John Wiley and Sons; New York: 1978; 316.
4. Foote, C.S. In Rånby, B., Rabek, J.F., eds. *Singlet Oxygen: Reactions with Organic Compounds and Polymers.* John Wiley and Sons; New York: 1978; 136.

PHOTOCHROMISM OF MERCURY DITHIZONATE

In sunlight, mercury dithizonate undergoes a photochemical isomerization reaction from an orange to a blue complex. The reversion of the photochemically produced blue isomer to the orange isomer is an example of a first-order reaction, which can be monitored using a red laser.

DEGREE OF DIFFICULTY

Experimental: difficult
Conceptual: moderate

MATERIALS

- red laser and/or green light source
- hot plate
- 3.1 g of mercuric acetate (Sigma product #M6279. Approx. $12 for 5 g or $27 for 100 g)
    or
- 2.2 g of palladium acetate (Sigma product #P6901. Approx. $93 for 5 g)
- chloroform
- laboratory oven
- beakers, Erlenmeyer flasks, test tubes, vial
- 300 mL of 6-M aqueous ammonia
- dithizone (diphenylthiocarbazone) (Sigma Product #D5130. Approx. $14 for 5 g)
- 10 g of phosphorous pentoxide or other suitable drying agent
- 10 mL of ethanol

PROCEDURE

1. Solutions of either mercury, palladium dithizonate, or both can be prepared using the instructions below. Note that the procedure for synthesizing the dithizonate complexes can be scaled down because only small quantities of the dithizonate complexes are needed to perform the experiment.

    a. In a fume hood, dissolve 5.0 g of dithizone (MW = 256.3) in 100 mL of hot 6-M aqueous ammonia.

    b. Dissolve 1.6 g of mercuric acetate (or 1.1 g of palladium acetate) in 50 mL of 6-M aqueous ammonia.

    c. Mix the solutions from steps a and b in a large beaker or Erlenmeyer flask, and gravity filter the resulting red–orange solid. The filtration takes several hours.

    d. Recrystallize the red cake from chloroform with the addition of a small amount of ethanol.

e. Dry the solid over phosphorous pentoxide (or other suitable drying agent) at 80°
   C for 3 hours (mp = 276° C).

f. In a fume hood, dissolve the mercury or palladium dithizonate prepared in steps
   a–e in chloroform to create solutions of approximately 10^{-5} M concentration.
   Fill a small screw-cap vial with some of the resulting solution and cap tightly.

2. Expose the mercury or palladium solutions prepared in step 1 to bright sunlight and
   observe any changes in color.

3. Remove the sample from bright sunlight and observe any color changes.

4. Repeat step 2 using a defocused red laser or green light source.

HAZARDS AND PRECAUTIONS

Mercury and palladium salts are toxic. Chloroform is a suspected carcinogen. Mercuric
acetate, phosphorus pentoxide, and ammonia are corrosives. Work in a fume hood and
wear gloves when preparing solutions and filling vials. Wear eye protection at all times.

DISPOSAL

Collect waste solutions in a waste bottle and dispose of in an approved manner. Mercury
and palladium wastes, as well as chloroform, may require special disposal procedures.

DISCUSSION

The term *photochromism* refers to the change of color of a substance under the influence
of light. Mercury dithizonate in the presence of sunlight produces a blue complex that
quickly reverts to the orange form upon standing in the shade. The chemical reaction
responsible for this change of color is believed to be the isomerization reaction shown
in Figure 6-5-1. The color changes in this reaction are pronounced and rapid when the

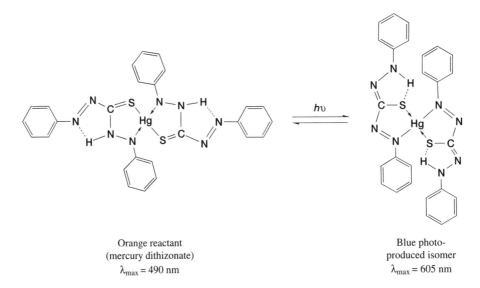

Orange reactant
(mercury dithizonate)
$\lambda_{max} = 490$ nm

Blue photo-
produced isomer
$\lambda_{max} = 605$ nm

FIGURE 6-5-1. Photoisomerization of mercury (II) dithizonate.

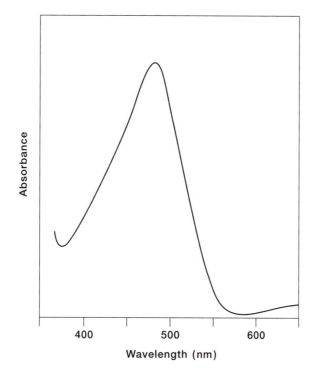

FIGURE 6-5-2. Absorption spectrum of mercury (II) dithizonate in xylene (1).

sample is placed in direct sunlight or intense artificial light in which wavelengths greater than 350 nm are present. This reaction is not stimulated by the red laser because the reactant does not absorb red light strongly (Figure 6-5-2).

The back reaction of the photochemically produced blue isomer to the orange form is rapid, which implies that the process of light activation is endoergic and that ample thermal energy is present at room temperature to overcome the energy barrier for the reverse reaction. Taking advantage of the fact that the blue isomer strongly absorbs red light but the orange isomer does not, Peterson and Harris (1) studied the kinetics of the back reaction by monitoring the absorbance at 605 nm with a recording spectrophotometer. Figure 6-5-3 shows their experimental results, which indicate that the reaction is first order in mercury dithizonate (1). The absorbance of light from a red He–Ne laser can also be used to monitor the reaction.

Palladium dithizonate has an absorbance maximum at approximately 630 nm. The substance is photochromic in sunlight and under irradiation of 532-nm light from a frequency-doubled yttrium–aluminum–garnet (YAG) laser, but low-power red and green helium–neon lasers cannot photochemically isomerize the complex. Apparently, the thermal back reaction is very rapid, and only powerful lasers can drive the reaction mixture toward the photoactivated complex to any appreciable extent.

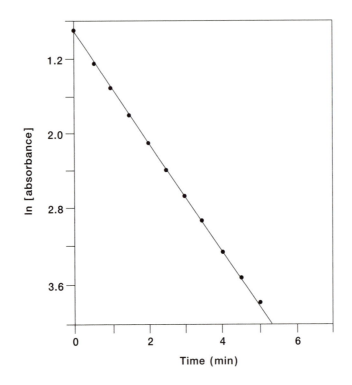

FIGURE 6-5-3. Kinetic data for the back reaction of photoactivated mercury (II) dithizonate (1). Absorbance readings were taken at 605 nm using a spectrophotometer.

REFERENCE

1. Peterson, R.L., Harris, G.L. *J. Chem. Ed.* 62 (1985) 802.

RUTHENIUM-CATALYZED PHOTOREDUCTION OF PARAQUAT

The herbicide paraquat acts as an electron acceptor in a photochemical reaction that is catalyzed by $Ru(bipy)_3^{+2}$ and that results in the oxidation of triethanolamine. A blue light source is used to drive the photochemical reaction, and a red laser is used to monitor the extent of the reaction.

DEGREE OF DIFFICULTY

Experimental: difficult
Conceptual: difficult

MATERIALS

Part 1: Preparation of [Ru(II)(bipy)$_3$] Cl$_2$·6H$_2$O, tris(2,2'-bipyridl) ruthenium (II) di-chloride

- mortar and pestle
- spatula
- balance
- oven
- protective gloves
- filter paper
- 10-mL beaker
- 100-mL beaker
- 100-mL round-bottom flask
- reflux condenser
- tubing to connect reflux condenser to cooling water
- stopcock grease to connect round-bottom flask to reflux condenser
- magnetic stir bar
- magnetic stir plate
- Büchner funnel with collection flask aspirator system
- calibrated 2-mL pipette plus bulb
- hot plate
- boiling chips
- 6 mL of hypophosphorous acid, 50%-by-weight solution (Aldrich product #21,490-6. Approx. $16 for 100 g)
- NaOH (Aldrich product #22,146-5. Approx. $14 for 25 g)
- 0.4 g of ruthenium (III) chloride (Sigma product #R8259. Approx. $48 for 5 g)
- 0.9 g of 2,2'-bipyridyl (Sigma product #D7505. Approx. $19 for 5 g)
- 12.6 g of potassium chloride (KCl) (Aldrich product # 22,146-5. Approx. $14 for 25 g)
- 40 mL of distilled water
- acetone
- ice bath

Part 2: Ruthenium-Catalyzed Photoreduction of Paraquat

- red laser[1]
- beam splitter
- light-detection system
- high-intensity light source with blue filter
- red–yellow cut-off filter (red and yellow plastic sheets)
- glass-sealing ampule, 1.0 mL
- pH meter
- torch
- 0.1 g of [Ru(bipy)$_3$] Cl$_2$ · 6H$_2$O (prepared in Part 1)
- 0.1 g of paraquat (methylviologin) (Sigma product #M2254. Approx. $8 for 250 mg)
- 1 mL of triethanolamine (Sigma product #T1377. Approx. $11 for 100 mL)
- nitrogen gas

PROCEDURE

Part 1: Preparation of [Ru(II)(bipy)$_3$] Cl$_2$·6H$_2$O

1. Grind hydrated ruthenium (III) chloride to a fine powder using a mortar and pestle.

2. Dry powder ruthenium (III) chloride in an oven at 120° C for at least 3 hours.

3. Prepare sodium dihydrogen hypophosphate (NaH$_2$PO$_2$) solution. This task is accomplished by placing a 10-mL beaker in an ice bath and slowly adding sodium hydroxide to 6 mL of 50%-by-weight hypophosphorous acid until a faint precipitate forms. This step is exothermic, so care must be taken to prevent runaway overheating. Use a calibrated 2-mL pipette for liquid transfer. Put 2 mL of the resulting solution into a 100-mL round-bottom flask.

4. Add 0.4 g of dry RuCl$_3$ and 0.9 g of 2,2'-bipyridyl to the 2 mL of NaH$_2$PO$_2$ solution in the round-bottom flask.

5. Add 40 mL of distilled water, making sure to wash down any material that has stuck to the neck or side of the round-bottom flask.

6. Place the magnetic stir bar in the round-bottom flask.

7. Equip the flask with a water-cooled reflux condenser. Use the stopcock grease to connect the round-bottom flask to the reflux condenser.

8. Using a heating mantle as a heat source and stirring with the magnetic stir bar, heat the round-bottom flask under reflux for 30 minutes. During this time, the color of the solution will change from greenish black to reddish orange. Then turn off the heating mantle and magnetic stir plate.

9. Wearing heat-resistant gloves, filter the hot solution into a 100-mL beaker. Do not let the solution cool before filtering it.

---

1. This experiment was tested using only a red He–Ne laser. Other red lasers may also give satisfactory results.

10. Add 12.6 g of potassium chloride to the filtrate. The clear filtrate should then appear cloudy. Add 2–4 boiling chips.

11. Heat the suspension to a boil on a hot plate to give a clear, deep-red solution. Make sure that no solid remains suspended in solution. Add distilled water as needed.

12. Wearing heat-resistant gloves, remove the boiling solution, which is completely clear, from the hot plate and place it on a paper towel on the bench top. Allow the solution to cool to room temperature. Then place the beaker in an ice bath to bring the remaining product out of solution.

13. Collect the $[Ru(bipy)_3]$ $Cl_2 \cdot 6H_2O$ product. Vacuum filter the contents of the beaker using a Büchner funnel. Use two 5-mL portions of cold (0° C) 10% acetone in water to wash out the beaker. Wash once with 30 mL of cold (0° C) 100% acetone. Cold acetone is a good washing agent because 2,2'-bipyridyl is highly soluble in cold acetone, whereas $[Ru(bipy_3)]$ $Cl_2 \cdot 6H_2O$ is not. Air dry the product.

Note: this synthesis is adapted from reference 1. It has been used successfully for the past 5 years in an introductory chemistry class at Stanford University for students with Advanced Placement credit in chemistry.

Part 2: Ruthenium-Catalyzed Photoreduction of Paraquat

Preparation of Solutions

1. Prepare 25 mL of an aqueous solution that is 10^{-5} M in $Ru(bipy)_3^{+2}$ and 0.2 M in triethanolamine (TEA) (MW = 149.2, density = 1.12 g/mL). The final pH should be approximately 7.5.

2. Prepare 25 mL of an aqueous solution that is 0.02 M in methylviologin (MW = 257.2). The final pH should be adjusted to between 7.0 and 7.5.

Preparation of Reaction Ampule

3. Wrap the bulb of a glass-sealing ampule with foil and fill it one-third full with the solution from step 1 and the same amount of the solution from step 2.

4. Degas the ampule by blowing a stream of nitrogen through a disposable pipette as shown in Figure 6-6-1. Try to prevent the solution from splashing up into the neck of the ampule.

5. Seal the ampule under a stream of nitrogen gas issuing from an inverted funnel (Figure 6-6-1).

Study of the Photochemical Reaction and the Thermal Back Reaction

6. Hold the ampule under bright sunlight or a high-intensity lamp and observe the color change. Remove the ampule from the strong lighting and allow the original color to return. Shaking the ampule accelerates the rate of the return reaction dramatically.

7. Set up an apparatus similar to that shown in Figure 6-6-2 to monitor light absorbance. Place a sample cell containing pure water in the apparatus and verify that the blue light source does not produce a change in light intensity at the probe detector.

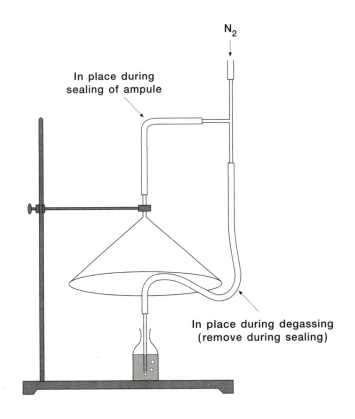

FIGURE 6-6-1. Apparatus used for degassing and sealing of ampule.

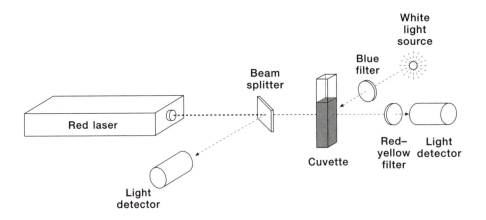

FIGURE 6-6-2. Schematic diagram of experimental setup for monitoring the photoreduction of paraquat.

8. Place the sample cell containing the reaction mixture in the apparatus. Irradiate the sample with the blue light source while monitoring the absorbance change with the red laser. Monitor the power output of the laser, and take measurements of the transmitted intensity only when the laser output is at some fixed value.

9. Turn off the blue light source when the absorbance reaches a maximum but continue to monitor the change in absorbance with the laser. If you are using a chart recorder, reduce the speed to a few centimeters per hour to minimize the use of chart paper. Record the transmitted intensity continuously as the solution slowly returns to its original color.

HAZARDS AND PRECAUTIONS

Ruthenium compounds are toxic. Triethanolamine is corrosive. Paraquat is a skin-absorbable poison. Wear gloves when handling all solutions, and wear eye protection at all times.

DISPOSAL

Collect waste solutions in a waste bottle and dispose of the bottle in an approved manner.

DISCUSSION

Paraquat (methylviologin) is a pesticide that is reduced in this experiment from a stable dicationic form (MV^{+2}) to a monocationic form (MV^{+1}) by an electron-transfer reaction with $Ru(bipy)_3^{+2}$. The structures of paraquat and $Ru(bipy)_3^{+2}$ are shown in Figure 6-6-3. The reaction of paraquat with ground-state $Ru(bipy)_3^{+2}$ is unfavorable, but the extent of the reaction can be increased by elevating $Ru(bipy)_3^{+2}$ to an excited triplet state (Figure 6-6-4), which is a better reducing agent than the ground state. In addition, the presence of triethanolamine at neutral pH can partially suppress the back reaction, because the TEA can react with the $Ru(bipy)_3^{+3}$ produced in the reaction, reducing it to $Ru(bipy)_3^{+2}$, which can then participate in the reaction cycle again. Thus, $Ru(bipy)_3^{+2}$ acts as a catalyst, but triethanolamine is consumed in the reaction; the TEA is first oxidized and then reacts with water to form various hydrolysis products (Figure 6-6-4). Despite the input of light energy and the use of triethanolamine, the reduction of paraquat (MV^{+2}) to MV^{+1} never goes to completion, in part because of several pathways by which the photochemically produced product (MV^{+1}) can be oxidized to the reactant.

The visible absorbance spectrum of the ground-state ruthenium complex is shown in Figure 6-6-5. Note that the complex absorbs strongly in the blue region but not in the red, and red laser light does not stimulate the reaction. Instead, blue light is used to drive the reaction, which is produced by filtering light from a high-intensity lamp through a blue plastic sheet that has the absorption spectrum shown in Figure 6-6-6.

The dicationic form of paraquat has an absorption maximum near 450 nm, which accounts for the original pale orange color of the solution. The reduced form of paraquat (MV^{+1}) is blue, and as the photochemical reaction progresses, the color of the solution

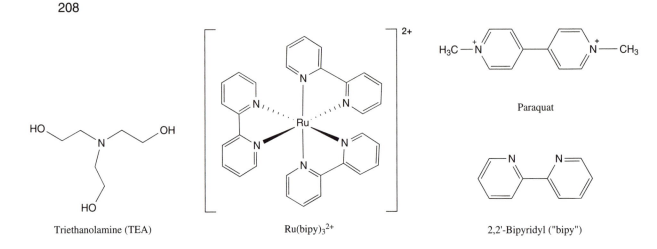

FIGURE 6-6-3. Molecular structures of important compounds involved in the photoreduction of paraquat.

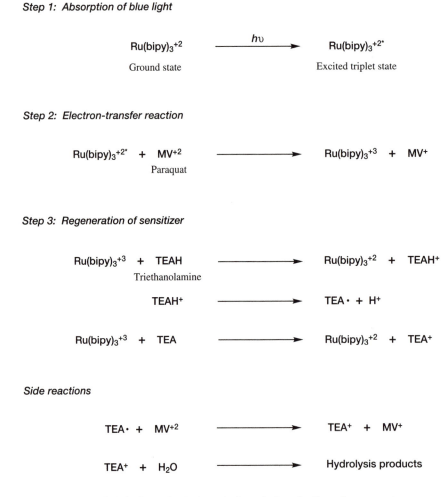

Step 1: Absorption of blue light

$$Ru(bipy)_3^{+2} \xrightarrow{h\upsilon} Ru(bipy)_3^{+2*}$$

Ground state   Excited triplet state

Step 2: Electron-transfer reaction

$$Ru(bipy)_3^{+2*} + MV^{+2} \longrightarrow Ru(bipy)_3^{+3} + MV^+$$
Paraquat

Step 3: Regeneration of sensitizer

$$Ru(bipy)_3^{+3} + TEAH \longrightarrow Ru(bipy)_3^{+2} + TEAH^+$$
Triethanolamine

$$TEAH^+ \longrightarrow TEA\cdot + H^+$$

$$Ru(bipy)_3^{+3} + TEA \longrightarrow Ru(bipy)_3^{+2} + TEA^+$$

Side reactions

$$TEA\cdot + MV^{+2} \longrightarrow TEA^+ + MV^+$$

$$TEA^+ + H_2O \longrightarrow \text{Hydrolysis products}$$

FIGURE 6-6-4. Important steps in the photoreduction of paraquat.

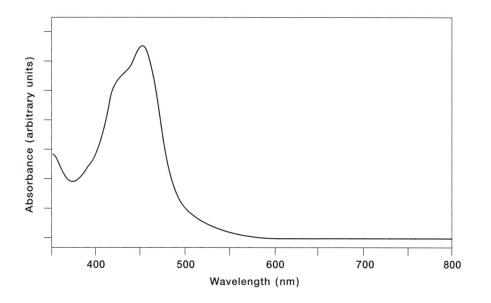

FIGURE 6-6-5. Absorption spectrum of $Ru(bipy)_3^{+2}$ in water (pH = 7.2).

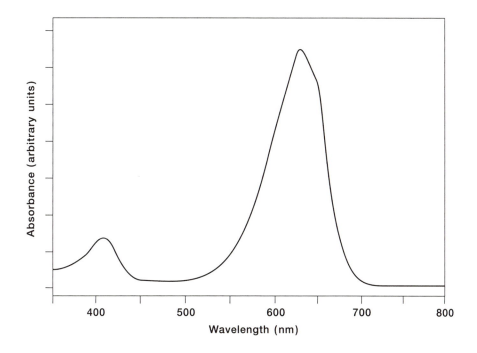

FIGURE 6-6-6. Absorption spectrum of blue plastic sheet used as a filter.

changes from orange–yellow to green and may even change to blue, if a low concentration of the ruthenium complex is used. Because the reduced form of paraquat is the only
species in solution that strongly absorbs red light, the extent of the reaction can be
monitored by charting the absorbance of red light as a function of time. To ensure that
the blue light used to drive the reaction is not detected along with the red light, the light
from the laser passes through a red–yellow cut-off filter before it enters the photometer.
The spectrum of the red–yellow filter (Figure 6-6-7) shows that blue light is effectively
blocked from entering the detector. A typical graph of absorbance vs. time for the photochemical reaction is shown in Figure 6-6-8.

The reaction is reversible, and the blue or green solution reverts to the original
orange–yellow color upon removal of the light source. The back reaction is much slower
than the photochemical forward reaction, as shown in Figure 6-6-9. The reverse reaction
is suppressed after a dozen or so reversible cycles, leading to a stable yellow–green
solution.

The reaction studied in this experiment is part of a somewhat more complicated
reaction cycle that was the focus of some early attempts to convert solar energy into
chemical energy. The key elements of this reaction cycle are given in Figure 6-6-10.
Note that the oxidation of paraquat is coupled to the production of hydrogen gas, which
is catalyzed by an Adam's catalyst (PtO_2), and that cysteine is substituted for TEA. The
presence of cysteine totally suppresses the back reaction of MV^{+2} at alkaline pH. The
overall reaction involves the reduction of water to hydrogen gas, which can be stored
for later use as a chemical energy source.

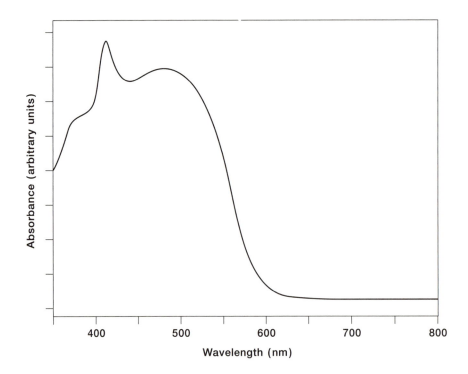

FIGURE 6-6-7. Absorption spectrum of red–yellow filter composed of red and yellow
plastic sheets.

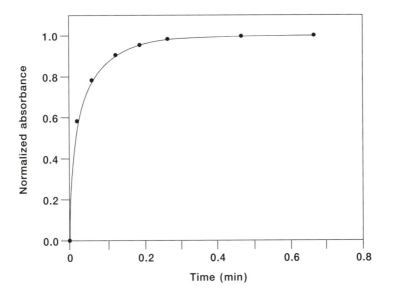

FIGURE 6-6-8. Kinetic data for the photoreduction of paraquat.

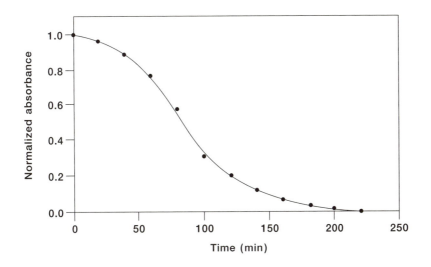

FIGURE 6-6-9. Kinetic data for the nonphotochemical back reaction of photoreduced paraquat to paraquat.

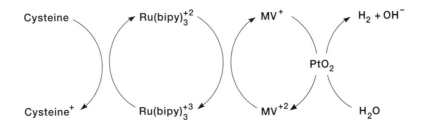

FIGURE 6-6-10. Reaction cycle involved in attempts to harness photochemical energy using a ruthenium-catalyzed reduction of paraquat.

REFERENCE

1. Broomhead, J.A., Young, C.G. *Inorg. Synth.* 21 (1982) 127.

Chemical Safety

This appendix is extensive, and its proper use requires understanding of the following key:

KEY TO NATIONAL FIRE PROTECTION ASSOCIATION (NFPA) HAZARD CLASSIFICATION

Health : Fire : Reactivity (instability) : *Other Properties*   (The numbers are in the
(0–4) : (0–4) : (0–4) : (*Oxidizer = Oxid; Water reactive = W̶*) order of increasing
                                                              severity.)

KEY TO RECOMMENDED HANDLING PROCEDURES

A	= Keep away from food/drink	**F,G,H**	= Protective clothing, rubber apron, rubber safety shoes
B/C	= Hood/respirator		
D	= Chemically resistant gloves	**I**	= Eye bath
E	= Safety goggles	**J**	= Safety shower

KEY TO RECOMMENDED CHEMICALLY RESISTANT GLOVE SELECTION[a]

1 = Neoprene 6780   **5** = Nitty Gritty 65NFW (Natural Rubber)
2 = Ultraflex Neoprene   **6** = Hustler 725R (PUC)
3 = Nitrosolve 727   **7** = Best Butyl 878 (Butyl)
4 = Ultraflexnitrile 22R   **8** = Best Viton 890 (Viton)

The above glove reference numbers follow the symbol **D** used in the *KEY TO RECOMMENDED HANDLING PROCEDURES*. The absence of these reference numbers for the remainder of the chemicals indicates that the above gloves have not been tested for those chemicals. The manufacturer's (reference 6) recommendation for laboratory work (dexterity/puncture/chemical-resistance requirements) is the type "N-Dex 8005." The recommendation for heat-resistant gloves is the type "6781 R/Neoprene."

[a] In cases where the chemical resistance of glove has not been tested, only the symbol D appears in Table A-1.

The National Fire Protection Association (NFPA) has developed a hazard classification system that appears as a diamond-shaped warning symbol on chemical containers. Figure A-1 explains this warning symbol.

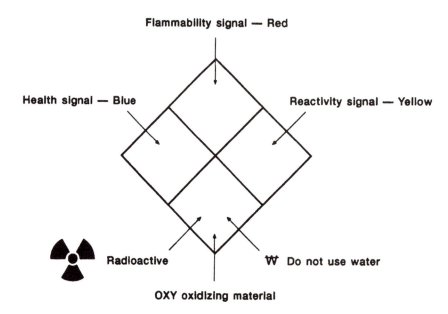

FIGURE A-1. The diamond-shaped warning symbol used for hazard classification of chemicals. The top, left, and right boxes refer to flammability, health, and reactivity hazards, respectively, and contain a number rating from 0 to 4. The bottom box is used for special hazards; the most common is a warning against the use of water, as indicated by the symbol W.

We also list useful references that explain the labeling requirements for the storage, transportation, and disposal of chemicals:

1. *Fire Protection Guide on Hazardous Materials,* 9th ed. National Fire Protection Association; 1 Batterymarch Park, Quincy, MA 02269. 1986.

2. *NFPA 49 Hazardous Chemical Data.* National Fire Protection Association; 1 Batterymarch Park, Quincy, MA 02269. 1991.

3. "Hazardous Materials Table," *Department of Transportation Code of Federal Regulations* 49, Part 172.101. 1977.

4. "California Hazardous Waste Codes Numbers." *Barclays California Code of Regulations,* Title 22, Division 4.5, Chapter 10, Article 5, Section 66261.126, Appendix XII. 1991.

5. "Federal EPA Hazardous Waste Code Numbers." *Code of Federal Regulations,* 40 CFR Part 261, Subpart D. 1993.

6. Best Manufacturing Co., Edison St., Menlo, GA 30731 (supplier of chemically resistant gloves).

We conclude this appendix by listing by experiment the chemicals used, their NFPA classifications, their health hazards, and their recommended handling procedures (Table A-1). We list the *first appearance* of a chemical to avoid repetition.

TABLE A-1. CLASSIFICATION, HAZARDS AND RECOMMENDED HANDLING OF
CHEMICALS USED IN THIS BOOK

Experiment No.	Compound	NFPA Hazard Classification	Health Hazard	Recommended Handling
2-1	Sulfuric acid (3 M)	3:0:2:—	Toxic; corrosive	
			Harmful through:	
			oral ingestion	A
			inhalation	B/C
			dermal absorption	D(1–8),E,F,G,H,I,J
	Sodium thiosulfate (3 M)	—:—:—:—	Irritant	
			May be harmful through:	
			oral ingestion	A
			inhalation	B/C
			dermal absorption	D(1–8)
2-3	Albumin	—:—:—:—	Not listed	
	Latex	—:—:—:—	Not listed	
2-4	Sodium chloride (saturated aqueous solution)	—:—:—:—	Irritant (mild)	
			May be harmful through:	
			oral ingestion	A
			inhalation	B/C
			dermal absorption	D
2-5	Potassium hydroxide (2 M)	3:0:1:—	Toxic; corrosive	
			Harmful through:	
			oral ingestion	A
			inhalation	N/A
			dermal absorption	D(1–8),E,F,G,H,I,J
	Oleic acid	0:1:0:—	Flammable; irritant	
			Harmful through:	
			oral ingestion	A
			inhalation	N/A
			dermal absorption	D(1–8),E,G,I
	Myristic acid	—:—:—:—	Irritant	
			Harmful through:	
			oral ingestion	A
			inhalation	B/C
			dermal absorption	D(3),E,I,J

(continued)

216

TABLE A-1. (continued)

Experiment No.	Compound	NFPA Hazard Classification	Health Hazard	Recommended Handling
2-5	n-Dodecane	0:2:0:—	Flammable; irritant (mild)	
			May be harmful through:	
			oral ingestion	A
			inhalation	B/C
			dermal absorption	D(3,7),E,I,J
	1-Pentanol	1:3:0:—	Flammable; irritant	
			Harmful through:	
			oral ingestion	A
			inhalation	B/C
			dermal absorption	D(1–8),E,F,I,J
	Hydrochloric acid (1 M)	3:0:0:—	Toxic; corrosive; reactive	
			Harmful through:	
			oral ingestion	A
			inhalation	B/C
			dermal absorption	D(1–8),E,F,I,J
2-6	Gelatin	—:—:—:—	Not listed	
	Sodium chloride (0.15-M aqueous solution)	—:—:—:—	Irritant (mild)	
			May be harmful through:	
			oral ingestion	A
			inhalation	B/C
			dermal absorption	D(1–8)
2-7	Sodium hydroxide (1 M)	3:0:1:—	Toxic; corrosive	
			Harmful through:	
			oral ingestion	A
			inhalation	B/C
			dermal absorption	D(1–8),E,F,G,H,I,J
2-8	Quinine sulfate	—:—:—:—	Toxic; irritant	
			Harmful through:	
			oral ingestion	A
			inhalation	B/C
			dermal absorption	D,E
	Glacial acetic acid	2:2:1:—	Flammable; toxic; corrosive	
			Harmful through:	
			oral ingestion	A
			inhalation	B/C
			dermal absorption	D(1,2,5,7),E,F,I,J

TABLE A-1. (*continued*)

Experiment No.	Compound	NFPA Hazard Classification	Health Hazard	Recommended Handling
2-8	Sucrose	—:—:—:—	Not listed	
2-9	Dimyristoyl phosphatidyl-choline	—:—:—:—	Irritant	
			Harmful through:	
			oral ingestion	A
			inhalation	B/C
			dermal absorption	D,E,F,I,J
	Sodium chloride (solid)	—:—:—:—	Irritant (mild)	
			May be harmful through:	
			oral ingestion	A
			inhalation	B/C
			dermal absorption	D(1–8)
	Potassium chloride	—:—:—:—	Irritant	
			Harmful through:	
			oral ingestion	A
			inhalation	B/C
			dermal absorption	D(1–8),E,I,J
	Disodium hydrogen phosphate	1:0:1:—	Irritant	
			Harmful through:	
			oral ingestion	A
			inhalation	B/C
			dermal absorption	D,E,I,J
	Sodium dihydrogen phosphate	1:0:1:—	Irritant	
			Harmful through:	
			oral ingestion	A
			inhalation	B/C
			dermal absorption	D,E,I,J
	Magnesium sulphate (anhydrous)	—:—:—:—	Irritant	
			Harmful through:	
			Oral ingestion	A
			inhalation	B/C
			dermal absorption	D,E,I,J
	Glucose	—:—:—:—	Not listed	
3-3	Glycerol	1:1:0:—	Flammable; irritant	
			Harmful through:	
			oral ingestion	A
			inhalation	B/C
			dermal absorption	D(1–8),E,I,J

(*continued*)

TABLE A-1. (*continued*)

Experiment No.	Compound	NFPA Hazard Classification	Health Hazard	Recommended Handling
3-3	Ethylene glycol	1:1:0:—	Flammable; toxic; irritant	
			Harmful through:	
			oral ingestion	A
			inhalation	B/C
			dermal absorption	D(1–8),E,I,J
3-4	Ion-exchange resin AG501-X8LD	—:—:—:—	Not listed	
4-1	Heptane	1:3:0:—	Flammable; irritant	
			Harmful through:	
			oral ingestion	A
			inhalation	B/C
			dermal absorption	D(8),E,F,I,J
	Silica gel	—:—:—:—	Irritant	
			Harmful through:	
			inhalation	B/C
			dermal absorption	D(1–8)
	Methanol	1:3:0:—	Flammable; toxic; irritant	
			Harmful through:	
			oral ingestion	A
			inhalation	B/C
			dermal absorption	D(3,4),E,F,I,J
	Naphthalene	2:2:0:—	Flammable; irritant	
			Harmful through:	
			oral ingestion	A
			inhalation	B/C
			dermal absorption	D,E,I,J
	o-Dichloro-benzene	2:2:0:—	Flammable; toxic; irritant	
			Harmful through:	
			oral ingestion	A
			inhalation	B/C
			dermal absorption	D(8),E,F,I,J
	Benzoic acid	2:1:—:—	Flammable; irritant	
			May be harmful through:	
			oral ingestion	A
			inhalation	B/C
			dermal absorption	D,F

TABLE A-1. (*continued*)

Experiment No.	Compound	NFPA Hazard Classification	Health Hazard	Recommended Handling
4-1	Methylbenzene (toluene)	2:3:0:—	Flammable; irritant	
			Harmful through:	
			oral ingestion	A
			inhalation	B/C
			dermal absorption	D(8),E,I,J
	Ethanol (ethyl alcohol)	0:3:0:—	Flammable; irritant	
			Harmful through:	
			oral ingestion	A
			inhalation	B/C
			dermal absorption	D(1,4),E,F,I,J
4-2	Limonene	—:—:—:—	Irritant	
			Harmful through:	
			oral ingestion	A
			inhalation	B/C
			dermal absorption	D(3,8),E,F,I,J
4-3	Salicylic acid	0:1:0:—	Flammable; toxic; irritant	
			Harmful through:	
			oral ingestion	A
			inhalation	B/C
			dermal absorption	D(1–8),E,F,G,I,J
	Dichloromethane (methlyene chloride)	2:1:0:—	Flammable; toxic	
			Harmful through:	
			oral ingestion	A
			inhalation	B/C
			dermal absorption	D(8),E,F,I,J
	Sodium bicarbonate	—:—:—:—	Not listed	
	Diethyl ether	2:4:1:—	Flammable; toxic; irritant	
			Harmful through:	
			oral ingestion	A
			inhalation	B/C
			dermal absorption	D(3,6),E,F,I,J
4-4	Glycidol	—:—:—:—	Toxic; irritant	
			Harmful through:	
			oral ingestion	A
			inhalation	B/C
			dermal absorption	D,E,F,I,J

(*continued*)

TABLE A-1. (continued)

Experiment No.	Compound	NFPA Hazard Classification	Health Hazard	Recommended Handling
4-4	Perchloric acid >50%	3:0:3:Oxid	Corrosive; irritant	
			Harmful through: oral ingestion inhalation dermal absorption	A B/C D,E,G,H,I,J
5-1	Aluminum oxide (alumina)	0:1:1:—	Flammable; irritant	
			Harmful through: oral ingestion inhalation dermal absorption	A B/C D(1,3,5),E,I,J
5-2	Chlorophyll solutions	—:—:—:—	Not listed	
5-3	Sodium iodide	—:—:—:—	Irritant	
			Harmful through: oral ingestion inhalation dermal absorption	A B/C D,E,F,I,J
	p-Benzoquinone	—:—:—:—	Highly toxic; irritant	
			Harmful through: oral ingestion inhalation dermal absorption	A B/C D,E,F,G,H,I,J
	Acetone	1:3:0:—	Flammable; irritant	
			Harmful through: oral ingestion inhalation dermal absorption	A B/C D(7),E,F,G,I,J
	1,2 Propanediol (propylene glycol)	0:1:0:—	Flammable	
			May be harmful through: oral ingestion inhalation dermal absorption	A B/C D(1,4,6–8)
5-5	Zinc chloride	—:—:—:—	Toxic; irritant	
			Harmful through: oral ingestion inhalation dermal absorption	A B/C D(1–8),E,F,I,J

TABLE A-1. *(continued)*

Experiment No.	Compound	NFPA Hazard Classification	Health Hazard	Recommended Handling
5-5	Tin (II) chloride	—:—:—:—	Toxic; corrosive	
			Harmful through:	
			oral ingestion	A
			inhalation	B/C
			dermal absorption	D(1–8),E,F,I,J
	Copper (II) bromide	—:—:—:—	Toxic; irritant	
			Harmful through:	
			oral ingestion	A
			inhalation	B/C
			dermal absorption	D,E,F,I,J
	Copper (II) sulfate pentahydrate	—:—:—:—	Toxic; irritant (severe)	
			Harmful through:	
			oral ingestion	A
			inhalation	B/C
			dermal absorption	D,E,I,J
	Nickle (II) nitrate hexahydrate	1:0:0:Oxid	Toxic; irritant	
			Harmful through:	
			oral ingestion	A
			inhalation	B/C
			dermal absorption	D,E,F,I,J
	Chromium (III) chloride hexahydrate	—:—:—:—	Irritant; severe	
			Harmful through:	
			oral ingestion	A
			inhalation	B/C
			dermal absorption	D,E,F,G,H,I,J
	Cobalt (II) chloride hexahydrate	—:—:—:—	Toxic; irritant	
			Harmful through:	
			oral ingestion	A
			inhalation	B/C
			dermal absorption	D,E,F,I,J
	Cobalt (II) acetate tetrahydrate	—:—:—:—	Irritant	
			Harmful through:	
			oral ingestion	A
			inhalation	B/C
			dermal absorption	D,E,F,I,J
	Iron (II) sulfate heptahydrate	—:—:—:—	Irritant	
			Harmful through:	
			oral ingestion	A
			inhalation	B/C
			dermal absorption	D,E,F,I,J

(continued)

TABLE A-1. *(continued)*

Experiment No.	Compound	NFPA Hazard Classification	Health Hazard	Recommended Handling
5-5	Potassium ferrocyanide trihydrate	—:—:—:—	Irritant	
			Harmful through:	
			oral ingestion	A
			inhalation	B/C
			dermal absorption	D,E,F,I,J
	Potassium ferricyanide	—:—:—:—	Irritant	
			Harmful through:	
			oral ingestion	A
			inhalation	B/C
			dermal absorption	D,E,F,I,J
	Nickel (II) hexamine bromide	—:—:—:—	Toxic; irritant	
			Harmful through:	
			oral ingestion	A
			inhalation	B/C
			dermal absorption	D,E,F,I,J
	Potassium chromium (II) acetate trihydrate	—:—:—:—	Toxic	
			Harmful through:	
			oral ingestion	A
			inhalation	B/C
			dermal absorption	D,E,F,I,J
	Iron (II) ammonium sulfate hexahydrate	—:—:—:—	Irritant	
			Harmful through:	
			oral ingestion	A
			inhalation	B/C
			dermal absorption	D,E,F,I,J
6-1	Methylene blue	—:—:—:—	Irritant	
			Harmful through:	
			oral ingestion	A
			inhalation	B/C
			dermal absorption	D,E,I,J
	Triethylamine	2:3:0:—	Flammable; toxic; corrosive; lachrimator	
			Harmful through:	
			oral ingestion	A
			inhalation	B/C
			dermal absorption	D(3),E,F,I,J

TABLE A-1. (*continued*)

Experiment No.	Compound	NFPA Hazard Classification	Health Hazard	Recommended Handling
6-2	*cis*-Dimethyl-maleate	—:—:—:—	Toxic; irritant	
			Harmful through:	
			oral ingestion	A
			inhalation	B/C
			dermal absorption	D,E,F,I,J
	Carbon tetrachloride	3:0:0:—	Toxic; irritant	
			Harmful through:	
			oral ingestion	A
			inhalation	B/C
			dermal absorption	D(3,8),E,F,I,J
	Bromine in carbon tetrachloride (0.6 M)	4:0:0:Oxid	Toxic; irritant	
			Harmful through:	
			oral ingestion	A
			inhalation	B/C
			dermal absorption	D(3,8),E,F,I,J
	Iodine in carbon tetrachloride (0.6 M)	3:0:0:—	Toxic; irritant	
			Harmful through:	
			oral ingestion	A
			inhalation	B/C
			dermal absorption	D(3,8),E,F,I,J
6-3	Acrylamide	4:0:0:Oxid	Toxic; irritant	
			Harmful through:	
			oral ingestion	A
			inhalation	B/C
			dermal absorption	D,E,F,I,J
	Triethanolamine	1:1:1:—	Flammable; irritant	
			Harmful through:	
			oral ingestion	A
			inhalation	B/C
			dermal absorption	D(1–8),E,F,I,J
6-4	1,3-Diphenyliso-benzofuran	—:—:—:—	Irritant	
			Harmful through:	
			oral ingestion	A
			inhalation	B/C
			dermal absorption	D,E,F,I,J

(*continued*)

TABLE A-1. *(continued)*

Experiment No.	Compound	NFPA Hazard Classification	Health Hazard	Recommended Handling
6-5	Chloroform	2:0:0:—	Toxic; irritant	
			Harmful through:	
			oral ingestion	A
			inhalation	B/C
			dermal absorption	D(8),E,F,I,J
	Aqueous ammonia (6 M)	—:—:—:—	Toxic; corrosive	
			Harmful through:	
			oral ingestion	A
			inhalation	B/C
			dermal absorption	D(1–8),E,F,I,J
	Diphenyl-thiocarbazone (dithizone)	—:—:—:—	Irritant	
			Harmful through:	
			oral ingestion	A
			inhalation	B/C
			dermal absorption	D,E,I,J
	Mercury (II) acetate	—:—:—:—	Highly toxic; corrosive	
			Harmful through:	
			oral ingestion	A
			inhalation	B/C
			dermal absorption	D,E,F,I,J
	Palladium (II) acetate	—:—:—:—	Irritant	
			Harmful through:	
			oral ingestion	A
			inhalation	B/C
			dermal absorption	D,E,I,J
	Phosphorous pentoxide	—:—:—:—	Highly toxic; corrosive	
			Harmful through:	
			oral ingestion	A
			inhalation	B/C
			dermal absorption	D,E,F,I,J
6-6	Tris(2,2'-bipyridyl) ruthenium (II) chloride hexahydrate	—:—:—:—	Toxic; irritant	
			Harmful through:	
			oral ingestion	A
			inhalation	B/C
			dermal absorption	D,E,F,I,J

TABLE A-1. (*continued*)

Experiment No.	Compound	NFPA Hazard Classification	Health Hazard	Recommended Handling
6-6	1,1'-Dimethyl-4,4'-bipyridinium dichloride (methyl viologin, or paraquat)	—:—:—:—	Highly toxic; extremely hazardous; corrosive	
			Harmful through:	
			oral ingestion	A
			inhalation	B/C
			dermal absorption	D,E,F,G,H,I,J
	Nitrogen gas	—:—:—:—	Suffocant	
			Harmful through:	
			inhalation	B/C
	Ruthenium (III) chloride	—:—:—:—	Corrosive	
			Harmful through:	
			oral ingestion	A
			inhalation	B/C
			dermal absorption	D,E,F,I,J
	2,2'-Bipyridyl	—:—:—:—	Toxic; irritant	
			Harmful through:	
			oral ingestion	A
			inhalation	B/C
			dermal absorption	D,E,F,I,J
	Hypophosphorous acid	—:—:—:—	Corrosive; irritant	
			Harmful through:	
			oral ingestion	A
			inhalation	B/C
			dermal absorption	D,E,F,I,J

INDEX